The Language Gap

About the Authors and Respondent

CLIFFORD A. WILSON Clifford Wilson holds the Ph.D. degree in psycholinguistics from the University of South Carolina and is the former Senior Lecturer in Education at Monash University in Melbourne (lecturer-in-charge of psycholinguistics). He currently serves as president of the Pacific College of Graduate Studies in Melbourne, Australia.

While Asssistant Professor of Early Childhood Education at the University of South Carolina (Aiken), Dr. Wilson compiled the Language Abilities Test, which examined the linguistic ability of children from early childhood to adulthood. This test was later developed into the *Language Abilities Guide*. Dr. Wilson has authored more than three dozen books.

He is a Fellow and Council Member of the Commercial Education Society of Australia and in 1971 was honored as "An Outstanding Educator of America." He is also a member of the American Psychological Association.

DONALD W. MCKEON Donald McKeon earned the Ph.D. degree in linguistics from New York University in 1972 and since 1973 has served as Assistant Professor of English and Linguistics at Virginia Polytechnic Institute and State University in Blacksburg, Virginia. Concurrently, Dr. McKeon coordinates the English as a Second Language Program at VPI & SU with continuing advising duties in the International Student office of the VPI & SU Graduate School.

Dr. McKeon is the author of *Modifying English as a Second Language*. He coedited (with Robert M. Grindell) *Intensive Course in American English Grammar* for the American Language Institute at New York University.

He is a member of the Linguistic Society of America, Teachers of English to Speakers of Other Languages, and the American Scientific Affiliation.

MARVIN K. MAYERS Marvin K. Mayers holds the following degrees: B.A., Wheaton College; M.Div., Fuller Theological Seminary; and Ph.D., University of Chicago. He now serves as Dean of the Graduate School of Intercultural Studies and World Missions at Biola University in LaMirada, California.

Dr. Mayers was formerly Professor of Linguistics and Anthropology at the International Linguistic Center of the Wycliffe Bible Translators in Dallas, Texas, and was concurrently Adjunct Professor of Linguistics at the University of Texas at Arlington. He held those posts from 1972 to 1982. Before 1972 he was head of the Department of Sociology and Anthropology at Wheaton College, Wheaton, Illinois.

Dr. Mayers is the author of *Christianity Confronts Culture*. He is coauthor (with Stephen A. Grunlan) of *Cultural Anthropology: A Christian Perspective* and (with Robert Webber and Lawrence O. Richards) of *Reshaping Evangelical Higher Education*. In addition, he has published numerous scholarly articles.

The Language Gap

Clifford A. Wilson
and
Donald W. McKeon

with a response by
Marvin K. Mayers

ZONDERVAN PUBLISHING HOUSE
OF THE ZONDERVAN CORPORATION
GRAND RAPIDS, MICHIGAN 49506

PROBE MINISTRIES
INTERNATIONAL
DALLAS, TEXAS 75251

Copyright	© 1984 by Probe Ministries International

Library of Congress Cataloging in Publication Data	Wilson, Clifford A.
	  The language gap.
	
	  Includes bibliographical references.
	  1. Language and languages. 2. Animal communication.
	  I. McKeon, Donald W. II. Title.
	  P106.W593 1984 401 83-23269

ISBN	0-310-35771-3
Publisher	Zondervan Publishing House, 1415 Lake Drive, S.E., Grand Rapids, Michigan 49506
Rights	All rights reserved. No part of this publication may be reproduced in any form or by any means without the prior permission of the copyright owner.
Place of Printing	*Printed in the United States of America*
Series Editor	Steven W. Webb, Probe Ministries
Design	Inside cover design by Paul Lewis Cover design by Paul Lewis Book design by Louise Bauer 84 85 86 87 88 89 - 10 9 8 7 6 5 4 3 2 1

What Is Probe?

Probe Ministries is a nonprofit corporation organized to provide perspective on the integration of the academic disciplines and historic Christianity. The members and associates of the Probe team are actively engaged in research as well as lecturing and interacting in thousands of university classrooms throughout the United States and Canada on topics and issues vital to the university student.

Christian Free University books should be ordered from Zondervan Publishing House (in the United Kingdom from The Paternoster Press), but further information about Probe's materials and ministries may be obtained by writing to Probe Ministries International, 12011 Coit Road, Suite 107, Dallas, Texas 75251.

Contents

TABLES, CHARTS, AND FIGURES   8

BOOK ABSTRACT   11

chapter one
IS THERE A LANGUAGE GAP?   13

chapter two
CHIMPS AND OTHER APES COMMUNICATE   19

chapter three
AN OVERVIEW OF HUMAN LANGUAGE   41

chapter four
A MODEL OF LINGUISTIC COMPETENCE   51

chapter five
LANGUAGE ACQUISITION: BIOLOGICAL BACKGROUND AND DEVELOPMENTAL STAGES   69

chapter six
THEORIES CONCERNING
LANGUAGE ACQUISITION   89

chapter seven
ANIMAL SYSTEMS OF
COMMUNICATION AND HUMAN
LANGUAGE   113

chapter eight
EVALUATION OF APE
COMMUNICATION WITH MAN   121

chapter nine
ACCOUNTING FOR THE LANGUAGE
GAP   155

RESPONSE   173

REFERENCES   179

FOR FURTHER READING   205

Tables, Charts, and Figures

Table 1
Signs Used by Washoe 24–25

Chart 1
Sarah's Chips and the Conditional Relation 27

Chart 2
Design Elements of Lana's "Yerkish" Keyboard 29

Table 2
Nim's Most Frequent Two- and Three-Sign Combinations 35

Table 3
Identification of Certain Semantic Relations in Koko's Sign Combinations 36

Table 4
Examples of Koko's Compound Names at Age 6+ 39

Figure 1
Three Complex Relations of Form and Meaning 47

Figure 2
A Transformational-Generative Grammar (Standard Theory) 54

Figure 3
Derivations for (5′) and (6) Paraphrase Set 55

Figure 4
Derivations for (5″) and (7) Paraphrase Set 56–57

Figure 5
Summary of How a Transformational-Generative Grammar Accounts for Ambiguity and Synonymy (Paraphrases) 58

Figure 6
Example of Deep Structure 63

Figure 7
Examples of Pivot/Open Words 81

Figure 8
Language Acquisition Model 99

Chart 3
Rhesus Monkey Calls 116

Figure 9
A Comparison of the Vocal Tract Configurations of Apes and Humans 117

Table 5
Koko's Expletives and Insulting Signs (From Age 4 to 7) 128

Table 6
Signs Koko Uses as "Insults, Expletives, and Derogations" 135

Figure 10
Double-blind Testing Apparatus Used in Project Washoe 143

Book Abstract

This book addresses the question of whether there is a qualitative difference between human language and animal systems of communication, especially in light of the recent attempts by psychologists to teach sign language and other signaling systems to apes. The major ape-man communication projects of the 1960s and 70s are described and evaluated with respect to what linguists have shown human language to be.

The knowledge of a native speaker, known as linguistic competence, is described as a highly complex system of rules that relates forms and meanings in an unlimited number of sentences. Given mere exposure to normal language use, all children acquire a fully developed system of competence within only a few years under greatly varying environmental conditions and individual differences. A comparison of two general theories of learning, empiricist and rationalist, reveals that only the latter with its emphasis on the contribution of an innately structured mind, can adequately account for the facts of universal language acquisition.

Compared with natural animal communication systems, human language is shown to be of a different order of complexity. Following a twelve-point characterization of human language structure and use, an evaluation of the primate learning experiments reveals that the apes do not demonstrate an ability to learn or use such a system; i.e., they cannot create sentences. As espoused by linguists, the rationalist view that human language acquisition is species-specific is thus closely scrutinized. Furthermore, the *more recent reappraisal by some of the leading ape communication researchers themselves only serves to further question the conclusions of those who believe the apes can learn humanlike signaling.*

In view of these findings, the question is raised as to how an evolutionary theory can account for this language gap. A review of three versions—continuity, discontinuity, and a hybrid of these—points out drawbacks in each case. The book concludes with a presentation of an alternative that incorporates a reasonable account of the language gap based on a difference in creation.

chapter one

Is There a Language Gap?

This chapter introduces the major questions covered in the book and explains why they are important. The now-famous ape language experiments have sought to demonstrate the primates' facility for acquiring language. The status of human linguistic ability is shown to be affected by the outcome of these experiments.

On the fifth of April, 1887, two months before her seventh birthday, Helen Keller experienced language contact for the first time in her frustrated life. Blind, unable to hear and unable to speak, she was, for the first few years of her life, almost void of communicative or educative experience. Then Anne Mansfield Sullivan came into her life.

Who would have thought, then, that this lonely and frustrated little girl would later be able to recount in these words her experience in meeting Miss Sullivan:

On the afternoon of the eventful day, I stood on the porch, dumb, expectant. I guessed vaguely from my mother's signs and

The Language Gap

from the hurrying to and fro in the house that something unusual was about to happen, so I went to the door and waited on the steps. The afternoon sun penetrated the mass of honeysuckle that covered the porch, and fell on my upturned face. My fingers lingered almost unconsciously on the familiar leaves and blossoms which had just come forth to greet the sweet southern spring. I did not know what the future held of marvel or surprise for me. Anger and bitterness had preyed upon me continually for weeks and a deep languor had succeeded this passionate struggle.[1]

The fact is that this eloquent and beautiful description of a spring day was written by a blind, deaf, and dumb human being. Furthermore, it was composed only fifteen years after that person had experienced her first real exposure to human language, the highest known level of communication.

What transpired in that short span of Helen's life that enabled her to move from frustration to eloquence in communicative ability? Anne Sullivan helped her to acquire language.

On the very first day Helen met her teacher, Miss Sullivan taught her to spell the word "d-o-l-l" by signs in the palm of her hand. Helen learned immediately how to spell the word, but by her own admission, had no idea what she was doing. "I did not know that I was spelling a word," she later wrote, "or even that words existed; I was simply making my fingers go in monkey-like imitation. In the days that followed I learned to spell in this uncomprehending way a great many words, among them *pin, hat, cup* and a few verbs like *sit, stand* and *walk.* But my teacher had been with me several weeks before I understood that everything has a name."[2]

Two of the words Miss Sullivan had tried in vain to get Helen to understand were "m-u-g" and "w-a-t-e-r." Finally, one day on a stroll down a familiar path she led Helen to a well where someone was drawing water. Placing one of Helen's hands under the gushing stream, Anne then spelled out w-a-t-e-r in the other hand, first slowly, then rapidly. Helen describes her emotions as the barrier to language broke: "I stood still, my whole attention fixed upon the motions of her fingers. Suddenly I felt a misty consciousness as of something forgotten—a thrill of returning thought; and somehow the mystery of language was revealed to me. I knew then that "w-a-t-e-r" meant the wonderful cool something that was flowing over my hand. That living word awakened my soul, gave it light, hope, joy, set it free! There were barriers still, it is true, but barriers that could in time be swept away."[3]

Indeed, that *was* just the beginning! Helen learned *thirty* symbols in that *one morning* when she grasped the fact that her teacher was making the sign for "water" on her hand.[4] Before this discovery she obviously had engaged in some kind of mental activity, but without a fully developed structure and set of categories provided by an acquired language for her thinking. Helen knew water, but she did not have the word or understand the symbol-referent relationship. But after the breakthrough, she demonstrated a remarkable facility for language acquisition, once the method of penetrating the barriers proved successful. Thirty words in a few short hours! It seemed as if an untapped capacity was waiting to be released once a nonauditory way of expressing herself was made available for her use.

Is There a Language Gap?

It would be difficult for us to imagine Helen Keller's dilemma before her acquisition of language—to think without words. Thought and language are difficult to separate. Yet we often cannot find the right word(s) to express feelings, ideas, intentions, or even to describe an event or situation; we scan our vast vocabulary to try to find that proper description for our experience. But without words we could hardly have begun the search. Such was Helen's dilemma.

Human language, traditionally considered to be the unique achievement and possession of human beings, is a symbolic, or representational system. The parameters of language—sound and meaning—bear no direct or simple relation with each other. The correspondence is arbitrary, unlike that between a signal (or sign) and its signification. Whereas a screech is a signal of fright, the sequence of the speech sounds "I'm scared!" symbolizes the emotional state for an

Thinking and Communicating Without Words

English speaker; "J'ai peur!" for a French speaker, etc. The famous American anthropologist-linguist Edward Sapir observed that this symbolic expressive power of language thereby makes it "a fit instrument for communication."[5]

These examples serve to show that the symbolism is complex as well. Not only are objects and concepts encoded, such as the speaker ("I" or "je"), but also more abstract relations of a predicate ("be scared" or "avoir peur"), which are formed with respect to the subject of the sentence. (The complexity and abstractness of the sound and meaning relationship will be a main theme in chapter 3.)

Before her experience with Miss Sullivan, Helen Keller was unable to communicate much beyond the simplest levels, her primary mode of communication being kinesic—body language (gestural and facial expressions), plus some audible, but nonverbal, expressions of frustration. She was limited, not because she could not think at all, but because she could not express her thoughts. She had no developed language. As the description of her first meeting with Miss Sullivan indicates, Helen possessed a world of thought behind those barriers. Once the barriers were removed, ever so slowly, she moved beyond the simple levels of communicating and began to express that world of her own thoughts through language.

Are There Crucial Differences?

The reality of the Helen Keller story points up some crucial questions related to the differences between animal communication systems and human language. The heart of the controversy centers on the observation that only humans seem to have such linguistic ability as a spontaneous and innate characteristic. Does this mean that there is an impregnable gap between human capacity for language and the communication systems of animals? Whether it does or not has some significant bearing on other questions, such as: Is man on an evolutionary continuum with the animal world? And, in light of the now famous ape/chimp language experiments, is man the exclusive possessor of language or not?

Furthermore, the answer to the language-gap question also has implications for the anthropologist and the issue of cultural influence. Many anthropologists today assume that culture precedes language. But does it? The difference between man and animal formerly was often illustrated by the fact that man uses tools—until researchers discovered animals that use tools, at least in a relatively primitive form. Today, many would hypothesize that the difference is determined solely by language. But if there is no language gap, is there *any* qualitative difference between man and animal? Further, while most researchers acknowledge there is *some* gap, how large a gap is it? Finally, if a gap does exist, is it a product of environment and cultural influence, or of innate mental structure, or a combination of these?

Is There a Language Gap?

Drs. E. Sue Savage-Rumbaugh and Duane M. Rumbaugh run the Yerkes primate research team in Atlanta, Georgia. The Yerkes team is one of the most prominent groups of researchers doing language experiments with the chimpanzee. Their experiments deal with the very *heart* of the language-gap question: Is man the only being capable of acquiring language?

They have elaborated the questions they are seeking to answer through their research in this way:

The Implications of a Difference in Man

> Are apes capable of language processes that are homologous with those of human beings, or is human language *unique?* What really is the essence of human language, and what, if anything, makes it *qualitatively* distinct from the natural communication system of chimpanzees? (italics added)[6]

The various experiments attempting to teach chimps sign language raise serious questions as to their implications concerning the uniqueness of human beings. The uninitiated may wonder about these points, but the chimpanzee researchers themselves believe these experiments have crucial significance relevant to the question. They expect to demonstrate, through intensive linguistic experimentation with chimps, gorillas, and related primates, the degree of kinship between the nonhuman primates and man, the nature of language, and new keys to human psychology. Thus the results or discoveries found through these experiments may have profound implications for us humans in the study of our identity and place in nature.

As we survey the studies of the researchers working with primates and compare their findings with what linguists tell us about human language, we will observe different assumptions about the nature of the object under study as well as of the theories of learning. We will also see that there are marked differences among the primate investigators as to their basic assumptions, methodologies, and conclusions. Not surprisingly, then, we will see that linguists themselves differ over the nature of certain aspects of language, owing to the complexity of those yet-to-be-understood areas on the frontiers of empirical investigation.

While this study certainly cannot be exhaustive, it is our hope that the material presented will enable the reader to achieve a better understanding of the issues and thus be in a better position to arrive at conclusions concerning the questions raised in this chapter.

chapter two

Chimps and Other Apes Communicate

The authors survey the major experiments on apes that have been attempts to teach these primates various forms of language signing. Reports cover early experiments on chimpanzees Gua and Viki, as well as the more recent efforts with Washoe, Sara, Lana, Nim Chimpsky, and Koko the gorilla.

Man's curiosity about himself, his world, and other inhabitants of his world has led him into innumerable forms of exploration and experiments. In recent years there have been many experiments with apes, thought by many to be man's closest relatives in the animal world. Because of the biological and anatomical similarities, much bio-medical research continues today with apes, as scientists search for solutions for man's physical problems and diseases. Since the early part of the twentieth century, psychologists have studied apes to understand man's psychological make-up, and, in particular, his acquisition and use of language.

Some Early Experiments

The first widely reported language experiments with apes involved attempts to make them speak verbally. The best known of these apes were the chimpanzees, Gua and Viki, both raised in the homes of the researchers alongside human infants. The early findings of the experiment with Gua were first published in 1933.

Gua, a female chimpanzee, was raised by Winthrop and Luella Kellogg.[7] The Kelloggs reared Gua with their own infant son, and reported that in sixteen months she learned to respond appropriately to about 100 words. Dr. Kellogg explained that "the scientific rationale for rearing an anthropoid ape in a human household is to find out just how far the ape can go in absorbing the civilizing influences of the environment. To what degree is the ape capable of responding like a child and to what degree will genetic factors limit its development?"[8] Eventually Gua responded to 166 words, but never showed any interest in producing speech sounds at all.

The most concerted attempt to teach speech to chimpanzees was that of Keith and Cathy Hayes with Viki.[9] Like Gua, Viki was sociable. Reared with a human infant, she seemed always ready to explore the human world. She even wanted to use some household appliances.

The Hayeses taught Viki to produce a sound on cue, "ah," a new sound not in the usual repertoire of chimpanzee signals. By holding her lips with their fingers the Hayeses taught her to utilize that sound and say something approximating "mama." She eventually learned to work her lips to make the appropriate sounds with manual help, but continued to use her own hands in the same way the Hayeses had done. In six-and-one-half years of persistent application by the Hayeses, Viki managed to mouth four near-words, these being "mama," "papa," "cup," and possibly "up." For each word, she produced the same vowel in a hoarse whisper. As Kellogg later wrote, the achievement by Viki represented the acme of chimpanzee achievement in relation to human speech.[10] Nevertheless, Viki's feeble attempts at speech-sound production were later explained by phonetician Philip Lieberman.[11] Lieberman has noted that while an ape has a larynx (with vocal cords) that *could* produce speech sounds, it does not have the same vocal tract configuration (above the larynx) as a person in order to do so.

Washoe, a Signing Chimpanzee

According to psychologists R. A. and B. T. Gardner, "The extent to which another species might be able to use human language is a classical problem in comparative psychology."[12] In some of the most famous chimp experiments ever conducted, the Gardners set out to teach the now well-known Washoe a nonvocal form of language, for by now it was recognized that apes could not really produce speech. In June 1966 the Gardners began their novel work with Washoe, who was about one year old and lived in a specially equipped trailer rather than in their home. Noises were allowed in her presence but not voiced speech; instead, the Gardners used American Sign Language (ASL), or Ameslan,[13] because of chimpanzees' ability to imitate and their natural manual dexterity.* Historically, chimpanzees involved with humans have been known to have a wide variety of communicative gestures, their manual anatomy being somewhat similar to that of humans. Moreover, the Gardners reasoned that Washoe's learning of ASL would allow for a comparative study with deaf children's acquisition of this system.[14]

The Gardners ensured that those who had exposure to Washoe communicated only by signing when in her presence. They shaped her hands into signs and

---

*See further in chapter 8 on the arbitrary (vs. iconic) nature of a sign language like ASL. Note that a sign language is not to be equated with finger spelling, although the latter may be used by a deaf person as a supplement to signing.

encouraged her to imitate them as well. They also used the technique of instrumental conditioning and found tickling to be the most effective incentive to communication.[15] Washoe stayed with the Gardners for four years before moving to the Institute for Primate Studies at the University of Oklahoma, where she was placed in the care of research associate Dr. Roger Fouts.

In determining Washoe's acquisition of signs, the Gardners set a fifteen-consecutive-day usage criterion of "appropriate and spontaneous occurrence."[16] On pages 24–25 we list some of the thirty signs that they state Washoe learned.

The Gardners reported that Washoe's two-sign combinations exhibited likenesses to children's two-word utterances, with respect to certain semantic categories (with some overlapping) identified by psycholinguist Roger Brown: appeal ("Please tickle"), location ("baby down"), action-object ("go flower"), agency ("you drink"), and attribution ("comb black").[17] Further, she could answer the sign of the question "What do you want?" by an appropriate sign-response such as "drink."

The researchers noted other interesting observations about Washoe's learning patterns, in addition to her multiple signing. As the experiment proceeded, they observed that Washoe transferred certain signs to other members of the same class of words; thus the word *dog* could apply to an unknown barking dog.[18] They also noted a simple form of categorizing, with errors usually involving another item in the same category, e.g., "brush" for "comb." Overall, however, the Gardners also found that Washoe often demonstrated "poor diction," nor could she always be relied on to make the correct response; when "pushed," Washoe would throw a tantrum.[19] Washoe's "manual diction" improved over a period of time, but the motivation was consistently to gain the offered reward, as can be seen from such signing sequences as "come-give-me," "more," "up," "sweet," "come tickle," and "food-eat."

After fifty-one months of training, Washoe was reported to have acquired 132 signs, using them as class referents, and to have formed many combinations expressing various semantic relations.[20] While the Gardners noted that Washoe's sign combinations gradually increased in length, they did not make distinctions in the order of her sign sequences, thus allowing for the attribution of different semantic interpretations. For example, "tickle me" and "me tickle" were both recorded as the former, Action-Object, rather than the latter, Agent-Action.[21]

More recently, Washoe became the subject of another important study at the Institute for Primate Studies.* She began "teaching" what she knew of sign language to her adopted son, Loulis. Dr. Fouts claimed at the time, "This will be the first case of cultural transmission of a language between generations of chimps, and it is going better than I expected." Loulis imitated Washoe's sign for "George"—one of her trainers—and used signs for "drink," "food," "hot," "fruit," and "give me."[22]†

Before making any conclusions concerning the cultural transmission of sign language among chimps, we must investigate the nature of what Washoe and other pongids have indeed acquired. And we must postpone any evaluation of chimps' attempts to learn a language system like ASL‡ until we consider the *nature* of *human* language.[23]

Chimps and Other Apes Communicate

Anne James Premack and David Premack began working with the chimpanzee Sarah in 1966, at the University of California at Santa Barbara.[24] These

Sarah and Artificial "Chip" Language

---

*One of the present authors (Dr. Wilson) was somewhat disappointed on a visit to Washoe at her home (then at the Institute for Primate Studies, Norman, Oklahoma). She did not make the signs as frequently as would be expected on the basis of some reports. Further, Washoe proved to be very much a chimpanzee, as demonstrated when she began screeching and making other normal animal sounds.

†Fouts has since left the Oklahoma Primate Center, along with Washoe and Loulis, and has started another primate research center at Central Washington University.

‡The Gardners have continued their sign training with some younger chimpanzees, alternating their approach somewhat, in accordance with what they had learned with pilot Project Washoe. More specifically, they exposed their subjects to signing from birth, and they used fluent signers from the start (see reference 23).

TABLE 1
Signs Used by Washoe Within the First 22 Months of Training (in order of appearance)

Signs	Description	Context
COME-GIMME	Beckoning motion, with wrist or knuckles as pivot.	Sign made to persons or animals, also for objects out of reach. Often combined: COME TICKLE, GIMME SWEET, etc.
MORE	Fingertips are brought together, usually overhead. (Correct ASL form: tips of tapered hand touch repeatedly.)	When asking for continuation or repetition of activities such as swinging or tickling, for second helpings of food, etc. Also used to ask for repetition of some performance, such as a somersault.
UP	Arm extends upward, and index finger may also point up.	Wants a lift to reach objects such as grapes on vine, or leaves; or wants to be placed on someone's shoulders; or wants to leave potty-chair.
SWEET	Index or index and second fingers touch tip of wagging tongue. (Correct ASL form: index and second fingers extended side by side.)	For dessert; used spontaneously at end of meal. Also, when asking for candy.
OPEN	Flat hands are placed side by side, palms down, then drawn apart while rotated to palms up.	At door of house, room, car, refrigerator, or cupboard; on containers such as jars; and on faucets.
TICKLE	The index finger of one hand is drawn across the back of the other hand. (Related to ASL TOUCH.)	For tickling or for chasing games.
GO	Opposite of COME-GIMME.	While walking hand-in-hand or riding on someone's shoulders. Washoe usually indicates the direction desired.
OUT	Curved hand grasps tapered hand; then tapered hand is withdrawn upward.	When passing through doorways; until recently, used for both "in" and "out." Also, when asking to be taken outdoors.

HURRY	Open hand is shaken at the wrist. (Correct ASL form: index and second fingers extended side by side.)	Often follows signs such as COME-GIMME, OUT, OPEN, and GO, particularly if there is a delay before Washoe is obeyed. Also, used while watching her meal being prepared.
HEAR-LISTEN	Index finger touches ear.	For loud or strange sounds: bells, car horns, sonic booms, etc. Also, for asking someone to hold a watch to her ear.
TOOTHBRUSH	Index finger is used as brush, to rub front teeth.	When Washoe has finished her meal, or at other times when shown a toothbrush.
DRINK	Thumb is extended from fisted hand and touches mouth.	For water, formula, soda pop, etc. For soda pop, often combined with SWEET.
HURT	Extended index fingers are jabbed toward each other. Can be used to indicate location of pain.	To indicate cuts and bruises on herself or on others. Can be elicited by red stains on a person's skin or by tears in clothing.
SORRY	Fisted hand clasps and unclasps a shoulder. (Correct ASL form: fisted hand is rubbed over heart with circular motion.)	After biting someone, or when someone has been hurt in another way (not necessarily by Washoe). When told to apologize for mischief.
FUNNY	Tip of index finger presses nose, and Washoe snorts. (Correct ASL form: index and second fingers used; no snort.)	When soliciting interaction play, and during games. Occasionally, when being pursued after mischief.
PLEASE	Open hand is drawn across chest. (Correct ASL form: fingertips used, and circular motion.)	When asking for objects and activities. Frequently combined: PLEASE GO; OUT, PLEASE; PLEASE DRINK.
FOOD-EAT	Several fingers of one hand are placed in mouth. (Correct ASL form: fingertips of tapered hand touch mouth repeatedly.)	During meals and preparation of meals.
FLOWER	Tip of index finger touches one or both nostrils. (Correct ASL form: tips of tapered hand touch first one nostril, then the other.)	For flowers.

Source: Reproduced from E. Klima and U. Bellugi, *The Signs of Language* (Cambridge, Mass.: Harvard University Press, 1979).

two psychologists were interested in the nature of brain mechanisms and in the relation between human language and animal systems. They questioned whether language was actually unique to the human species:

> It seems clear that language is a general system of which human language is a particular, albeit remarkably refined, form. Indeed, it is possible that certain features of human language that are considered to be uniquely human belong to the more general system, and that these features can be distinguished from those that are unique to the human information-processing regime. If, for example, an ape can be taught the rudiments of human language, it should clarify the dividing line between the general system and the human one.[25]

The Premacks stated that they were searching for Sarah's "underlying brain mechanisms," on the assumption that "to a large extent teaching language to an animal is simply mapping out the conceptual structures the animal already possesses."[26] Reward-reinforcement, therefore, was very important to the search, and the tasks that were required of Sarah to obtain the reward were graduated to become progressively harder. The Premacks started by using her existing patterns of behavior such as visual imitation and the use of gestures, and they capitalized on her manual dexterity. She had to exchange variously shaped and colored symbols for objects, with the reward given only when the correct answer was given; e.g., a plastic square had to be placed on the language board instead of a banana, before the banana would be given as a reward. Then an apple replaced the banana, and soon the verb "give" was introduced with its plastic symbol.

Eventually, report the Premacks, Sarah was using 112 symbols, including 8 names of persons or chimpanzees, 21 verbs, 6 colors, 21 foods, 16 miscellaneous objects, and 30 "concepts, adjectives and adverbs."[27] She could identify colors, sizes, and shapes, based on a limited choice of two. By 1969 Sarah could use 130 colored plastic signs, with a reliability of about 75–80 percent, according to the authors.[28] "Apple" was a triangle, "banana" a square, "apricot" a rectangle, "Sarah" a seated monkey pictured from the back, "Mary" a fancy large M, with other signs for verbs as well as relations like "same as," "different from," and "if-then." See Chart 1 below.

As with the names of objects and colors, Sarah was trained to recognize the relational symbols through reward-reinforcement. The conditional relation "if-then" was taught with a single symbol based on the rewards following certain conditions; e.g., the trainer would give Sarah a choice of fruits—apple or ba-

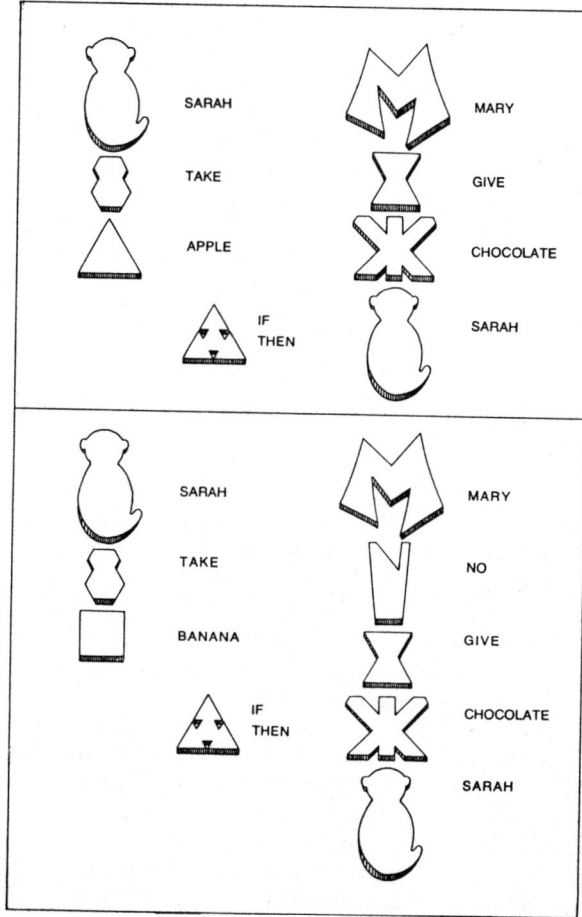

CHART 1.
Sara's Chips and the Conditional Relation
Source: A. J. Premack and D. Premack, "Teaching Language to an Ape," Scientific American 227 (October 1972): 97.

nana—and then a piece of chocolate only if she chose an apple. Then the conditional relation was presented graphically: "apple if-then chocolate, banana if-then no chocolate," etc. The Premacks report that while Sarah made many errors at first, she eventually learned the correct usage of the "if-then" symbol.[29]

The question must be asked whether these activities qualify as instances of language usage, especially in light of the heavy conditioning and limited choices involved.[30] Survival demands that wild animals have these *stimulus-response* abilities, albeit in a different form.

While the Gardners did not distinguish between Washoe's different order of signing, the Premacks reinforced English word order by rewarding Sarah's production of "give apple," but not "apple give."[31] More importantly, however, is the Premacks' claim that in successfully carrying out the command, Sarah comprehended the abstract sentence structure in "Sarah insert apple pail banana dish" (elliptical for "Sarah insert apple [in] pail" and "Sarah insert banana [in] dish"). They reason that Sarah would have had to know "that 'apple' and 'pail' go together but not 'pail' and 'banana,' even though the terms appear side by side . . . that the verb 'insert' is at a higher level of organization and refers to both 'apple' and 'banana' . . . (and) that she, as the head noun, must carry out all the actions."[32]

Lana's Language Machine

Incorporating certain elements of the Premacks' approach, Duane Rumbaugh has led a research team to teach chimpanzees to use a specially designed computer. He started this work in 1970 at Yerkes Regional Primate Research Center in Atlanta, Georgia, with the young chimpanzee Lana.[33] According to Rumbaugh, "Our ultimate goal is to better understand the etiology [the study of cause or causation] of language development in man; our immediate goal is to determine unequivocally the anthropoid's capacity for linguistic production including conversation."[34]

With the system of communication being computer-based, Lana had access to a large keyboard displaying lexigrams on lighted keys; such word symbols were the vocabulary of an artificial language, "Yerkish," with its specifications of lexigram sequences for correct "sentences."[35] Each lexigram was a distinctive geometric white symbol on a colored background. See Chart 2, below.

At first only relevant keys were lighted, and pressing only one key would bring the desired reward. However, the tasks became increasingly complex as Lana was gradually trained to press multiple keys in the correct order, e.g., "Please Machine give M&M period."[36]

The locations of the symbols on the keyboard were altered from time to time in order to determine that it was genuine knowledge of the visual symbol, and not

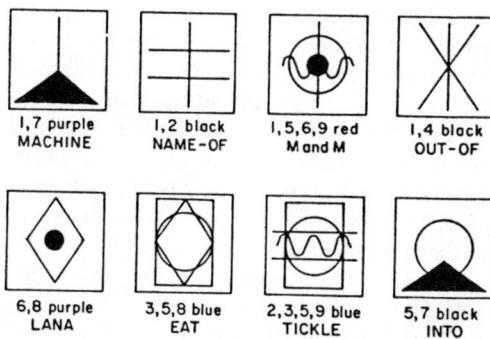

CHART 2.
Design Elements of Lana's "Yerkish" Keyboard
Source: Ernst Von Glasersfeld, "The Yerkish Language and its Automatic Parser," in D. M. Rumbaugh, ed. Language Learning by a Chimpanzee: THE LANA PROJECT (New York: Academic, 1977), p. 94.

spatial memory, that Lana was demonstrating. The symbols themselves were changed from time to time; according to experiments, about fifty were available at any one time. As more sophisticated equipment was utilized, the number of lexigrams was increased.

Early in the project Lana was considered a "polite" chimpanzee: she pressed the key for "please" before making her request. But actually she had been trained to do this so that the machine would know that a request was forthcoming. In fact, it would not give her what she asked for unless she started out with "please," ended with "period," and used an acceptable sequence of words. Lana was said to have learned to press "period" if she made a mistake; on the assumption that she knew that there was no point in continuing, she would start again in the only way that would be reward-producing—by complying with the machine's program requirements.

It is important to notice that reward-reinforcement was used to an extreme with Lana: by pressing certain keys in the correct order she could get such rewards as food, drink, music, trips to the monkey playground, films, or a window opened to view outside activities.

Using an extension of the language-training system designed for Lana, and working with four young chimpanzees from the Yerkes nursery, Rumbaugh and his colleagues later set out to understand (in a more precise way than had been possible from earlier investigations)[37] the processes involved in the initial acquisition of language skills.

The experimenters kept very detailed records of their work and the histories of each of the four young chimps: Ericka, Sherman, Austin, and Kenton. All but Erica were born at Yerkes. The four chimps lived together in the language-project quarters.

The first of the language experiments involved the attempt to teach them to label objects, according to either generic (color), or specific (objects/names) aspects. Over a four-month period none of the chimpanzees succeeded in using the lexigram symbols as labels or names (i.e., with a "semantic connection between lexigram and referent").[38] According to Savage-Rumbaugh and Rumbaugh, "What this phase of the work did show was that the labeling or naming of objects or attributes of objects is not the way in which to begin language training with a nonlinguistic organism such as the chimpanzee."[39] Such were the conclusions concerning Experiment 1—the chimps did not achieve *true* labeling, i.e., as linguistic expressions, or *words*.

Experiment 2 further seemed to indicate that the chimps responded to certain lexigrams as conditional stimuli, not "words" to be used for requesting the item available."[40] In Experiment 3, the researchers taught the chimps to string together food names with the appropriate verbs. The chimps did learn to string together sequences (e.g., "give orange") to gain the described results. However, tests with Sherman, the most advanced chimp among them, "indicated that any comprehension of the semantic content was meager. Without a set toward a particular sentence, performance decreased remarkably."[41]

The Rumbaughs concluded that it was not until Experiment 4 in the series that any semblance of languagelike performance began to appear. It was at this point that words previously used in Experiments 1 and 2 ("juice," "M&M," "sweet potato,") were reintroduced. The chimps were given the opportunity to use the skills they had learned in Experiment 3 in a primarily stimulus-response manner.[42]

At the end of this particular study (which is not the final work to be done with these particular chimpanzees), the animals had a working vocabulary of fourteen words. However, the researchers felt confident of the level of symbolization demonstrated by the chimps, because the chimps' use of the lexigrams was not strictly dependent on contextual cues.[43] The Rumbaughs thus offer a strong caution concerning any hasty assignment of "wordness" to an ape's response by an investigator, remarking that "it must not be presumed that this word functions for the chimpanzee in a manner similar to that in which a word functions for human beings.[44]

The researchers conclude:

> Most important, though, we believe that the present research has demonstrated that even the simple symbolic skills such as those

required by the animals in the present study, if semantically based, conceptually sound, and acquired to a high degree of accuracy, provide for a type of conceptual reorganization and restructuring of information that is uniquely symbol-dependent and which increases the animal's conceptual understanding and awareness of the world around him.[45]

Thus questions concerning the symbolic status of these and other pongids' behaviors have been raised by some of the researchers themselves. This issue will be further discussed after considering symbolism in human language.

A Chimp With a Challenging Name: Nim Chimpsky

While the Rumbaughs began to show concern regarding chimps' learning of true word symbols, a Columbia University psychology professor, Herbert Terrace, had questions of his own concerning the earlier language-related experiments. He writes in the beginning of his noted book *NIM: A Chimpanzee Who Learned Sign Language:*[46]

> Having read the publications of the Gardners, Premack and Rumbaugh, I was convinced that the state of the art of teaching chimpanzees to use language had advanced considerably by using non-vocal languages compatible with the physical nature of a chimpanzee. There was an abundance of evidence showing that a chimpanzee could learn to acquire a respectable vocabulary of words whose form had no obvious relationship to the objects they symbolized. I remained skeptical, however, about the evidence the Gardners, Premack, and Rumbaugh presented that implied that a chimpanzee could create sentences or that their motivation to use language was sufficient to allow them to engage in conversations about things other than their basic needs. At the same time the achievements of Washoe, Sara, and Lana, the high degree of intelligence in chimpanzees, and the degree to which humans seemed to be able to figure out their feelings and moods (and vice versa) gave me hope that chimpanzees could be taught to use language in a humanlike manner. That hope prompted me to start Project Nim.[47]

Thus, with the desire to determine whether chimps could create sequences exhibiting true sentential properties, Terrace commenced his work in late 1973 with an infant chimp, whom he named "Neam Chimpsky" or "Nim" for short, after the noted linguist Noam Chomsky. Terrace, a former student of the well-known behavioral psychologist B. F. Skinner, became interested in Chomsky's claims concerning the language and mind, in contrast to Skinner's views (see chapter 6). Briefly, Chomsky has strongly maintained that human language is innately specified in its essential characteristics and is *qualitatively different* from animal communication systems.[48] Skinner, on the other hand, has viewed language as conditioned behavior, describable in terms of external, environmental factors that shape the learning.[49]

Terrace had Nim raised as though he were a human child and, like the Gardners with Washoe, had him trained to respond in sign language through molding his hands. Terrace and his numerous project workers kept extensive records of Nim's progress and made some videotapes as well. Terrace established two criteria for judging Nim's sign learning: (1) three independent observers had to report the "spontaneous occurrence" of a sign (not immediately preceded by a teacher's sign) and (2) the sign had to be recorded as "spontaneous" and "appropriate" on five successive days.[50]

Nim's teachers recorded more than 19,000 combinations of two or more signs from the ages of nineteen to thirty-nine months. They whispered into a small tape recorder important information about the sign and later transcribed the message in a more detailed report.[51] A combination of signs was determined to be any sequence not interrupted by the return of the hands to resting position.[52] In almost four years of the project, Nim was observed to express 125 signs and comprehend some 200 signs. Of the 19,000 plus combinations (*tokens*), Terrace and his associates analyzed over 5,000 *types* of two or more combinations,[53] with a high percentage of a relatively few words occurring in the sequences, as seen in the 25 most frequent two- and three-sign sequences. See Table 2, page 35.[54]

These sequences were recorded in the order of signing, unlike the Gardners' data for Washoe. In assessing all the data after Project Nim was completed, Terrace observes that in the last two years there was no increase in the length or complexity of Nim's utterances,[55] and that the longer ones were quite repetitive;[56] more specifically, the three-sign utterances were no more informative than the two-sign ones; e.g., *play me* versus *play me Nim*.[57]

Thus the most comprehensive data collection of all the signing apes revealed a definite leveling off of Nim's ability, much to Terrace's disappointment.

Koko: A Gorilla Gets Into the Act

Chimps are not the only subjects of language-learning experiments. The well-publicized achievements of Koko the gorilla have also gained the attention of primatologists, the media, and thus the public. Francine Patterson of Stanford University has been working with Koko since 1972, making use of ASL and spoken language to communicate. In prefacing her own work, Patterson agrees that more questions than answers have surfaced in the man-ape communication so far, and she sees the need for "additional work with the great apes, more carefully and clearly delineating the similarities and differences between the acquisition and use of language by apes and humans."[58]

Koko has been taught signs by the same method that the Gardners used with Washoe, and Terrace with Nim, molding the hands into a sign representing an activity or an object while in its presence. Assuming that "vocabulary development is one of the best indexes of human intelligence," Patterson happily reports the following:

> Koko's vocabulary grew at a remarkable pace. Over the first year and a half, she acquired about one new sign every month. After 36 months of training, Koko was reliably using 184 signs—that is, she used each spontaneously at least once a day, 15 days out of a month. By age 4½, she had 222 signs by the same criterion. By 6½, she had used 645 different signs [at least once]. . . . I would estimate that Koko's current working vocabulary—signs she uses regularly and appropriately—stands at about 375.[59]

Patterson suggests that Koko is learning language in the same way that human infants do: in the early

TABLE 2
Nim's Most Frequent Two- and Three-Sign Combinations

Two-Sign Combinations	Frequency	Three-Sign Combinations	Frequency
play me	375	play me Nim	81
me Nim	328	eat me Nim	48
tickle me	316	eat Nim eat	46
eat Nim	302	tickle me Nim	44
more eat	287	grape eat Nim	37
me eat	237	banana Nim eat	33
Nim eat	209	Nim me eat	27
finish hug	187	banana eat Nim	26
drink Nim	143	eat me eat	22
more tickle	136	me Nim eat	21
sorry hug	123	hug me Nim	20
tickle Nim	107	yogurt Nim eat	20
hug Nim	106	me more eat	19
more drink	99	more eat Nim	19
eat drink	98	finish hug Nim	18
banana me	97	banana me eat	17
Nim me	89	Nim eat Nim	17
sweet Nim	85	tickle me tickle	17
me play	81	apple me eat	15
gum eat	79	eat Nim me	15
tea drink	77	give me eat	15
grape eat	74	nut Nim nut	15
hug me	74	drink me Nim	14
banana Nim	73	hug Nim hug	14
in pants	70	play me play	14
		sweet Nim sweet	14

Source: Herbert S. Terrace, Nim: A Chimpanzee Who Learned Sign Language (New York: Knopf, 1979), p. 317.

stages she engaged in "manual babbling . . . playfully experimenting with gestures she has been taught."[60]

Patterson states that Koko, like children, "has spontaneously generalized and overgeneralized her signs to novel objects and situations," citing, for example, how she applied the sign for "straw" to label plastic tubing, cigarettes, and a car radio antenna.[61]

Moreover, Koko is reported to have formed signs (that resembled ASL signs), though Patterson remarks that they "may be natural gorilla gestures."[62] Other innovative gestures not seen (in untrained gorillas) turn out to be iconic, i.e., resembling what is signified.[63]

Concerning sign combinations, Patterson observes that Koko has formed very "meaningful and sometimes novel strings of two or more," starting with "gimme food" at about fourteen months.[64] She reports that Koko's average length of signed utterance (see chapter 5, page 77, for *M*ean *L*ength of *U*tterance) is about 2.7 signs, as compared with Nim's 1.5 signs per utterance.[65]

Patterson analyzes these sign combinations in eleven semantic categories used by psycholinguist Roger Brown for children most of which are indicated in the following table:[66]

TABLE 3
Identification of Certain Semantic Relations in Koko's Sign Combinations

Brown's Semantic Relations	Koko's Signs
Nomination (Identification)	that bird
Attribution (Noun Modification)	hot potato
Possession	Koko purse
Agent-Action	you eat
Action-Object	catch me
Locative (Place Reference)	go bed
Dative (Indirect Object)	give me drink

After Francine Patterson, "The Gestures of a Gorilla: Language Acquisition in Another Pongid," *Brain and Language*, vol. 5, no. 1 (January 1978): 90 (Table 2, "Early Semantic Relations Expressed by the Gorilla Koko, The Chimpanzee Washoe, and Deaf Children"). These data were collected by Patterson during Koko's third year. (See chapter 4, the Semantic Representation section concerning the semantic, or thematic, relations; also some examples of semantic relations exhibited in the speech of a two-year-old child in chapter 5, the Two-Word Stage section [pp. 80–83].)

Yet Patterson regrets not having done a full study of Koko's word order to date (1981).[67]

Patterson claims that Koko modulates her signs, varying the motion of a sign to indicate a specific actor, or she will "selectively modify one of the formational components—the configuration of the hand, motion of the sign, or place where the sign is made—while holding the others to standard form."[68] Based on this and other attributions of Koko's inventiveness, Patterson concludes:

> It suggests that Koko has a grasp of the underlying structure from which signs are generated, and indicates that Koko uses Ameslan in the "open" and "creative" way that Terrace thought was critical if an ape was to be credited with language.[69]

In other interesting claims, Patterson reports that Koko "tells" her when she feels happy or sad, and, with a near-human sense of humor, she "insults" her human companions, and "lies" to them on occasion if by doing so she can "avoid blame."[70] Furthermore, Patterson indicates Koko has demonstrated the possibility of an embryonic understanding of time displacement, referring to both past and future events. Once when told to drink, she made the signs for "Later Koko drink."[71]

Patterson has also tested Koko on the Stanford-Binet Intelligence Scale, with scores reportedly ranging between 84 and 95 according to human IQ measurement. If indeed this is an accurate assessment, it is remarkable, especially if cultural biases toward humans will necessarily show up when such a test is given to a gorilla.[72] A good example of the cultural difference would be that when asked on the test where he or she would run to shelter from the rain—the choices being a hat, a spoon, a tree, or a house—a child would choose the house, whereas Koko chose the tree. Good enough for a gorilla, of course, but because according to human standards the correct answer would be the house, Koko had to be marked wrong.

A further interesting point is that Koko has been reported to respond to things that her trainer says in English by translating them into signs. One example occurred when Patterson asked, "Do you want a taste of butter?" Koko signed, "Taste butter."[73]

In another recent development, Professor Patrick Suppes and his colleagues at the Stanford Institute for Mathematical Studies have designed a keyboard computer that "enables Koko to talk by pressing buttons linked to a speech synthesizer."[74] The keyboard has forty-six buttons, each representing a word with an arbitrary geometric shape in one of ten colors. Koko's requests, like "Want apple eat," are being recorded and stored in a computer data file for future analysis.[75]

More on Koko's "Creativity and Aptness of Expression"

Patterson also reports that Koko's production of signs is mostly "appropriate to the situation" and sometimes "strikingly apt," as seen in her labeling of a stale sweet roll as a "cookie rock."[76] Other examples are found in Table 4 (p. 39).

Patterson further claims Koko can rhyme, based on her auditory perception of spoken words. The first instance she records occurred when Koko, having difficulty signing *need* signed *knee*—a near harmony—instead; at other times she interchanged *I* and *eye*, *know* and *no*, and *eleven* and *lemon*, etc.[77] On another occasion, Patterson reports the following: "When asked to identify broccoli once, she [Koko] signed *flower stink fruit pink . . . fruit pink stink*. I said, 'You're rhyming, neat!' Then to my astonishment, Koko signed, *Love meat sweet*."[78]

As another example, Patterson offers this exchange between associate Barbara Hiller (who is presumably speaking) and Koko in 1980, in a setting where some toy animals were placed in a row in front of Koko:[79]

Barbara: Which animal rhymes with hat?
Koko: *Cat*
Barbara: Which rhymes with big?
Koko: *Pig there.* (She points to the pig.)
Barbara: Which rhymes with hair?
Koko: *That.* (She points to the bear.)
Barbara: What is that?

Koko: *Pig cat.*
Barbara: Oh, come on.
Koko: *Bear hair.*
Barbara: Good girl. Which rhymes with goose?
Koko: *Think that.* (Points to the moose.)

Patterson's Claim: A Difference in Degree, Not in Kind

Patterson likens Koko's language development to Washoe's, stating that "both species of ape have exhibited close parallels to human children with respect to the development of semantic relations in early language.[80] While admitting there are differences between ape and human language acquisition, e.g., rate of acquisition, she declares that "the difference is one of degree and not of *kind*" (italics hers).[81]

Thus during the fifteen or so years that comparative psychologists have studied ape-human communication, perhaps no stronger claim or challenge to lin-

TABLE 4
Examples of Koko's Compound Names at Age 6+

Objects Referred to	Sign Combinations
tapioca pudding	milk candy
stereo viewer	mask look
frozen banana	fruit lollipop
tweezers	pick face
pomegranate seeds	fruit red seeds
parsley sprig	lettuce grass
mask	nose fake

Note that some of these could conceivably be classified in more than one way; e.g., "pick face" could be a "command" to do something with the tweezers.

After Francine Patterson and Eugene Linden, *The Education of Koko* (New York: Holt, Rinehart, and Winston, 1981), p. 147 (Table 5).

guistic research can be made than Patterson's conclusion: "Language is no longer the exclusive domain of man."[82]

What Patterson and other researchers who take such a radical stand are saying is that apes can truly symbolize in some nonvocal way. Further, these researchers claim that apes can creatively and appropriately combine such symbols to form sentences that are based on highly intricate principles, as we will see in the following chapters.

We now turn, then, to a linguistic characterization of human language and to an analysis of how children acquire it. After this discussion we will return to an evaluation of these claims concerning the relation of ape sign/visual symbol usage and human language.

chapter three

An Overview of Human Language

This chapter describes basic characteristics of human language, as found in every society. The relationship between sound (or form) and meaning is shown to be highly complex, governed by principles that native speakers have mastered but are essentially unaware of.

Human language is pervasive. Like air in the environment, it seems to permeate nearly every aspect of human life; indeed, without it there would be no rational thought, no communication of thought and feelings, and, ultimately, no human culture as we now know it. Wherever normal human life is found, there is language. Except for cases such as severe brain damage or other abnormalities, every child acquires the language of the speech community he or

The Pervasiveness of Language

she is raised in.* Such acquisition takes place within only a few years without any special instruction. Yet, as we will see, the knowledge of the language that is acquired continues to amaze and even at times baffle linguists in its complexity and scope. Despite the fact that hundreds of languages in the world are still without a writing system, linguists have emphatically maintained that there are no "simple" languages. Over sixty years ago, linguist Edward Sapir declared in his monograph *Language:*

> There is no more striking general fact about language than its universality. . . . The fundamental groundwork of language—the development of a clear-cut phonetic system, the specific association of speech elements with concepts, and the delicate provision for the formal expression of all manner of relations—all this meets us rigidly perfected and systematized in every language known to us. Many primitive languages have a formal richness, a latent luxuriance of expression, that eclipses anything known to the languages of modern civilization. Even in the mere matter of the inventory of speech the layman must be prepared for strange surprises. Popular statements as to the extreme poverty of expression to which primitive languages are doomed are simply myths.[84]

The Evasiveness of Language

Ironically, despite the advancements in modern science and technology and the dissemination of such information in this century, there is still an appalling lack of understanding among us concerning not only "exotic" languages but also our own native language. One often hears remarks like "I don't know any grammar," offered, perhaps, as an apology for the sentence forms one uses or for an inability to explain some rule of grammar. The truth of the matter is that one could not even form such a sentence without knowing the rules of the language—whether to form and arrange the words into a grammatical and meaningful sentence or even to pronounce the se-

---

*As we will discuss in chapter 5, language acquisition will be spontaneous in a child's early school years, but "the incidence of 'language-learning-block' rapidly increases after puberty" as noted neurologist Eric Lenneberg points out.[83]

quence of sounds. Underlying the actual production and interpretation of sentences we find a highly intricate system of principles that determine the relation between sound and meaning in a language. These principles, or rules, are internalized by a native speaker and may be called *linguistic competence,* following the description of the renowned linguist Noam Chomsky.[85]

If, then, a native speaker has this competence, or an "internalized grammar," why might one think that he/she does not know the grammar of the language? The answer is that the knowledge is subconscious. When we produce and interpret speech signals, we are not conscious of any system of principles, any more than we are aware of the intricate functioning of the respiratory system when breathing. Thus we would not expect one to give an accurate account of any of the principles which he/she subconsciously puts to use.[86]

An Overview of Human Language

To demonstrate that we make use of a system of rules in producing and understanding sentences, let's consider the very sentence that one could use to disavow such knowledge, and compare it with some other strings of words:
(1) I don't know any grammar.
(2) I not know any grammar.
(3) Know any don't grammar I.
(4) I don't know no grammar.

As native speakers of English, we would concur that (1) is grammatical, or well-formed, but would judge (2) and (3) to be deviant; (2) seems only slightly anomalous, such that we can still understand it, whereas (3) appears to be a jumble of words—totally ungrammatical. This range of deviance (degrees of grammaticality, according to Chomsky),[87] is attributable to the degree of violation of the rules that determine the formation of sentences; if the violation is minimal, one can probably still interpret the sentence; otherwise not.

Concerning (4), we observe that it is a paraphrase of (1), despite the two negatives; the only difference

The Organization of Language

in form is the occurrence of the negative *no* instead of the indefinite *any*. Although many claim that it is a "bad" sentence, linguists and sociolinguists are in agreement that sentence (4) demonstrates as much of the rule system of the language as sentence (1) does. As a matter of fact, historically, both single and double negatives were found in educated usage prior to the *prescriptive* approaches to English grammar prevalent in the eighteenth century. Thus a "rule" restricting common usage was invoked by Robert Lowth in his *Short Introduction to English Grammar* (1762), thereby prohibiting double negatives from use in "standard" dialects.[88] Nonstandard dialects have retained the double negative construction, which was once acceptable in all dialects in English.

It is important, therefore, to distinguish so-called "rules" of a *prescriptive* nature from those that are descriptive of one's subconscious knowledge of the language. Other prescriptive rules include restrictions against the use of split infinitives, object pronouns after the verb *be*, and *can* for *may* to express permission. Rules like the above are based on an erroneous assumption that those forms that are prevalent in nonstandard dialects as well as informal standard usage somehow "corrupt" the language. The split infinitive rule was inappropriately taken from Latin, which has one-word infinitives, hence impossible to "split"; object pronouns following *be* were found in standard English prior to the 1700s; *can*, like other auxiliaries, has undergone an extension in meaning, as part of the normal process of language change.[89]

If, then, we exclude certain prescriptive notions about how sentences "should" be formed, we would find our judgments concerning the grammaticality of strings like (1)–(4) to be quite uniform. This conclusion should not be surprising because sociolinguists like William Labov[90] have repeatedly emphasized that *all* dialects, including nonstandard ones, are systematic, i.e., regular. We can further argue that not only have we (native speakers) internalized a system of rules distinguishing grammatical from ungrammatical sentences but we have also internalized essentially the *same* system. Otherwise, our judgments would differ in unpredictable ways to the extent that we might not be able to communicate with one another.

To say that the system is essentially the same for all native speakers of a language does not rule out variations from dialect to dialect within a speech community. Nevertheless, many leading sociolinguists have taken the position that the differences between a nonstandard and a standard dialect are mainly on the "surface" level (in the sense that we describe on pages 59–61), as, for example, Labov[91] working with Northern urban Black English and Wolfram[92] with Southern rural Black English. Again, these and other sociolinguists have stressed the regularity as well as the essential similarity of all dialects.[93] To quote Labov, "In general, we find that non-standard English dialects* are not radically different systems from standard English but are instead closely related to it. These dialects show slightly different versions of the same rules, extending and modifying the grammatical processes which are common to all dialects of English."[95]

The Complexity of Language

Having introduced the notion that all native speakers subconsciously make use of a system of rules in producing and understanding sentences, we turn now to look at the nature of the relation between sound (or form) and meaning. Language use obviously involves discourse—connected speech and writing—beyond the individual sentence level. However, following the approach of transformational generative grammar as espoused by Noam Chomsky (see chapter 4), we will concentrate on the sound-meaning correspondence in sentences such as (5)–(10) below, recognizing that various aspects of meaning such as pronoun reference must take account of larger portions of the language.†

---

*The reader may find it worthwhile to look at a sociolinguistic study of Appalachian dialect, Wolfram and Christian (1976),[94] with respect to its systematic features and certain implications for teaching standard English to such nonstandard speakers.

†To justify this limitation of study would take us beyond the scope of the present work; see chapter 4, Semantic Representation, pages 61–67.

(5) The teacher has bored children.
(6) Children have been bored by the teacher.
(7) The teacher has children who are bored.
(8) It's possible to help the children.
(9) The children are possible to help.
(10) The children are likely to help.

First, we note that sentence (5) is ambiguous, having the meanings expressed in (6) and (7). (In this discussion, we take meaning to be limited to those aspects of the interpretation determined by the structure of the sentence, i.e., strict linguistic meaning, what Chomsky[96] describes as "logical form." See chapter 4.) Sentences (8) and (9), on the other hand, have the same meaning; they exhibit the relation of synonymy, or paraphrase.

Sentences (9) and (10) appear similar in form but are, upon reflection, quite different in meaning with respect to the *thematic* relations.[97] That is, the logical subject (or *agent*) of the verb *help* in (9) is not stated, while the logical object (or *patient*) of the verb is *the children*; i.e., someone may help the children. In contrast, these thematic relations are reversed in (10): the agent of the verb *help* is *the children,* while the patient is not stated; i.e., the children may help someone. This pair of sentences illustrates yet another relation exhibiting the complexity of the form-meaning relation in language; let's call it *opacity,* (from "opaque"); that is, the similarity in forms *conceals* the great difference in their respective meanings. In effect, we ignore the obvious analogy of forms in interpreting them.[98] There is further evidence that (9) and (10) are quite different: while (9) has (8) as a paraphrase, (10) does not have a corresponding paraphrase; (11) is ungrammatical (asterisked to show the deviance):

(11) *It's likely to help the children.

Therefore, these three relations—ambiguity, synonymy (or paraphrase), and opacity—demonstrate that form and meaning in human languages are not related in a simple, one-to-one way. We may schematically represent these complex relations as follows:

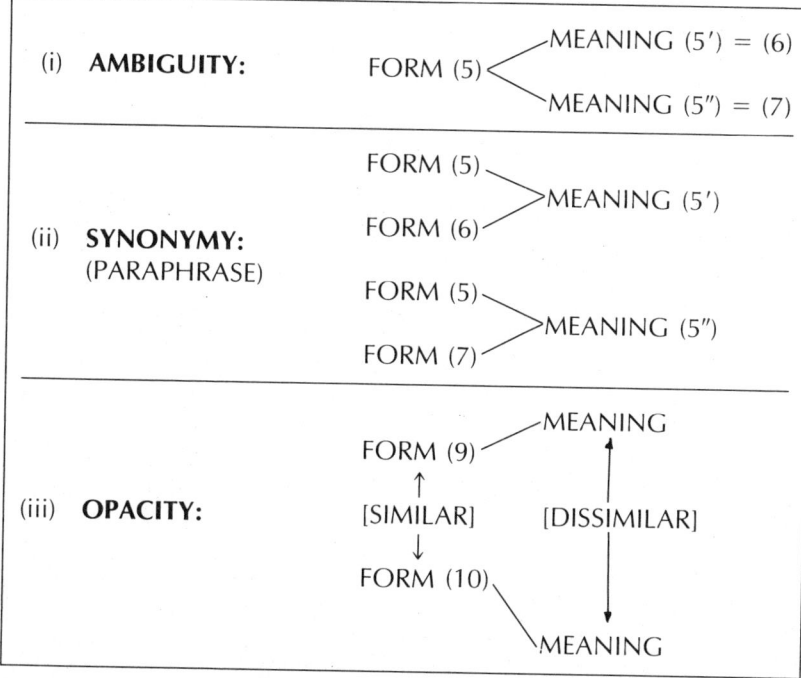

FIGURE 1. Three Complex Relations of Form and Meaning

If we attempted to list every word we knew, in principle we could do it. However, if we tried to list every sentence we knew, the task would never end; likewise for all the sentences we could understand. This ability to produce and understand an unlimited number of sentences has been called the "creative aspect of language use" by Chomsky[99] and described as innovative, stimulus free, and appropriate to the situation.

The Openendedness (Productivity) of Language

> Having mastered a language, one is able to understand an indefinite number of expressions that are new to one's experience, that bear no simple physical resemblance and are in no simple way analogous to the expressions that constitute one's linguistic experience; and one is able, with greater or less facility, to produce such expressions on an appropriate occasion, despite their novelty and independently of detectable stimulus configurations, and to be understood by others who share this still

mysterious ability. The normal use of language is, in this sense, a creative activity.[100]

It should be obvious that knowing a language cannot be equated with memorizing a stock of sentences and analogizing from them. Likewise, we must reject the idea that the size of one's vocabulary can restrict this productivity. Notice that any one sentence is potentially indefinite in length, not because of the extent of the vocabulary used, but because of the nature of the structure, as illustrated in the following:

(12) (a) I know that you know that the teacher knows that the children know . . . .
(b) This is the dog that bit the cat that chased the rat that ate the cheese that . . . .
(c) Because I know that you know that . . . , I will . . . .
(d) This is the dog that . . . , and that is the dogcatcher who . . . .

A common property of human languages is the potentiality of a sentence structure to be part of (*embedded* in) a larger sentence, as illustrated in (12 a–c). In traditional terms, such constructions are called subordinate clauses. In (12d) a coordination of sentences is illustrated. These and other properties of sentences are describable in terms of linguistic rules. Though the number of rules that we acquire is obviously limited, we are able to make use of such knowledge in unlimited and novel ways; human language is open-ended.

There is in principle, then, no linguistic limit either to the number of possible sentences in a language or to the length of any one sentence. Yet there are certain to be limiting factors in actual usage, such as memory limitations in processing sentences, expectations concerning appropriateness, and others.

Language Structure Versus Language Use

The last point touches on a theoretical distinction made in transformational grammar[101]—the distinction between linguistic competence and linguistic performance, i.e., between the system of implicit knowledge and the actual use of it. Speech production and interpretation involve other, nonlinguistic factors such as psychological state, memory, perceptual strategies, and motor coordination. Quite commonly, we make mistakes in speaking or listening because of the complex interplay of these components functioning in a particular social/environmental setting. For example, a memory limitation due to a state of tiredness may result in the production of a sentence fragment; noise in the environment may distract the hearer from correctly interpreting an otherwise perfectly understandable utterance.

Chomsky has thus proposed that the study of our knowledge of a language be approached ideally, that is, isolating grammatical competence from (though obviously interacting with) other mental structures and nonlinguistic knowledge that make up linguistic performance.[102] As psycholinguists Helen and Charles Cairns state, "It is theoretically necessary to separate, conceptually, the structure of knowledge and the use of knowledge. It is clear, however, that psycholinguistic, or performance, processes are far more complex and far more basic to linguistic behavior than they were once believed to be."[103]

The Psychological Reality of Language: Discreteness

In every aspect of language, linguists find evidence of the rule-governed relationship between sound and meaning. It has become clear, in fact, that the units of language we "hear" are a function of the rules we have internalized.[104] This discreteness of language—the separation of units of sound and words—is a perceptual phenomenon; that is, an interpretation of the continuum of sound signals imposed by the rules of the grammar. These units are not clearly discernible from an acoustic record of the sound waves produced.[105]

To support this conclusion, we turn to some data from a phonetic description of the sounds we make. We perceive words in sentences to be composed of individual sounds. For example, we would consider the word *strand* to be made up of six individual sound units (*s-t-r-a-n-d*). Yet, as an X-ray film of the vocal tract would reveal, the vocal organs are in constant motion—there are no breaks between the supposed sounds, nor do the vocal organs, such as the tongue, abruptly change from one position to another. Actually, there are over one hundred muscles in the vocal tract that are used in the articulation of speech sounds; consequently, several hundred individual muscle events may take place *every second*.[106] (See Figure 9, page 117.) In terms of an articulatory description of *strand*, we observe that the lips are partially rounded for the *s* segment in anticipation of the third segment, *r*, which is normally articulated with lip rounding. Compare the articulation of the *s* segment in *stand*, where no lip rounding occurs. This phenomenon of neighboring sounds influencing one another is called *assimilation*, various types of which are statable as phonological rules (see chapter 4, The Phonological Component . . . , pages 59–61). That we even regard these initial segments to be the same when, in fact, they are phonetically different, is further evidence of our rule-oriented perception of language.

We can better understand this conclusion concerning the influence of our internalized rules by considering how a totally strange language sounds to us. Instead of recognizing words, we do not "hear" much else than a stream of unintelligible signals principally because we do not know the rules of that language.

We conclude this chapter with the observation that language is a product of the human mind. It is the task of linguists to discover the intricate system of principles latent in the mind that constitute our knowledge of a particular language.

chapter four

A Model of Linguistic Competence

The relation between sound and meaning in language is further discussed. The rule systems of native speakers are complex and therefore linguists propose theories, or models, that help explain these highly complex cognitive systems. The authors especially point to a Transformational Generative Grammar model of linguistic competence.

As we have seen, knowledge of a language entails a subconscious mastery of a system of principles governing the formation and interpretation of sentences. Because the relation between sound and meaning is complex, we would expect this rule system to be intricate and abstract. In this chapter, we will consider the linguist's goal to make this intuitive knowledge explicit, i.e., to propose models or theories of competence.

The Language Gap

A Distinct Cognitive System of Linguistic Competence

There are various linguistic theories about a native speaker's knowledge of a language, depending on the assumptions one makes concerning what constitutes linguistic knowledge vs. nonlinguistic knowledge. The position we take here, following Chomsky, is that a distinct cognitive system of linguistic competence (grammatical knowledge)[107] interacts with systems of belief and knowledge of the world in determining the full meaning of utterances in context and interacts with other mental structures in actual linguistic performance, as previously mentioned in chapter 3. A theory of performance would include theories of both grammatical and *pragmatic* or *communicative competence*—that is, the knowledge of linguistic rules vs. the knowledge of the appropriateness of usage in specified cultural contexts (e.g., given various intentions of the speaker).[108] (See "Semantic Representation," pp. 61ff.)

Assuming this idealization of grammatical competence, Chomsky has compared the linguist's task of writing a grammar to that of a child acquiring his or her native language: The linguist's task is "to determine from the data of performance the underlying system of rules that has been mastered by the speaker-hearer and that he puts to use in actual performance."[109] The extent to which any such representation captures the generalizations of the language that the child has acquired enables linguists to compare their theories and the underlying assumptions.

The "Standard" Model and the Syntax of Sentences

In terms of a model put forward by Chomsky,[110] now known as the "standard" theory of transformational grammar,* there are three interacting subsystems: a syntactic component, a semantic compo-

---

*This model has served as a springboard for much research, not to mention debate, over the past twenty years. Many of the questions have concerned whether a distinction can be made between the linguistic meaning of an utterance and its full meaning in a performance context (see further discussion under "Semantic Representation," pp. 61ff.), and the role of deep structure with respect to the determination of meaning. In working with other linguists,

nent, and a phonological component—each with characteristic sets of rules. The syntactic component is viewed as central in that its rules *generate,* or specify, structures that are semantically and phonologically interpreted. Certain syntactic rules, called *phrase structure rules,* generate *deep structure phrase markers;* these are treelike configurations of phrases that determine the strict linguistic meaning of sentences. The other syntactic rules are called *transformations;* they generate *surface structure phrase markers* from deep structure phrase markers. Surface structures, in turn, are interpreted by phonological rules to specify the phonetic form of a sentence. Figure 2 on p. 54 is a schematic diagram of the standard model of linguistic competence. It is "generative" in that it contains a set of explicit rules that account for the form-meaning relation, the goal being to specify all of the grammatical, but not ungrammatical, sentences acceptable to a native speaker. The theory is "transformational" in that it contains transformational rules that relate deep and surface structures. The inclusion of this type of rule allows for the distinction in levels of structure so as to account for the observed complexity of the form-meaning relation.

In order to account for an ambiguous sentence like (5) in chapter 3—with meanings equivalent to (6) and (7)—the grammar would generate two different deep structures; different transformational rules would subsequently apply to the deep-structure phrase marker and successively derived phrase markers to yield two surface structures that do not reveal the deep-structural differences. Conversely, for paraphrases like (8) and (9) the grammar would generate one deep structure and, through the application of certain transformations, two different surface structures.

Chomsky himself has proposed substantial revisions in the "extended standard theory."[111] A fuller treatment of contemporary linguistic theory would necessitate discussion of these and other revisions, but the empirical issues involved would go beyond the scope of this present study. The standard theory, then, suffices to provide the reader with a general framework concerning a linguist's representation of native speaker competence.

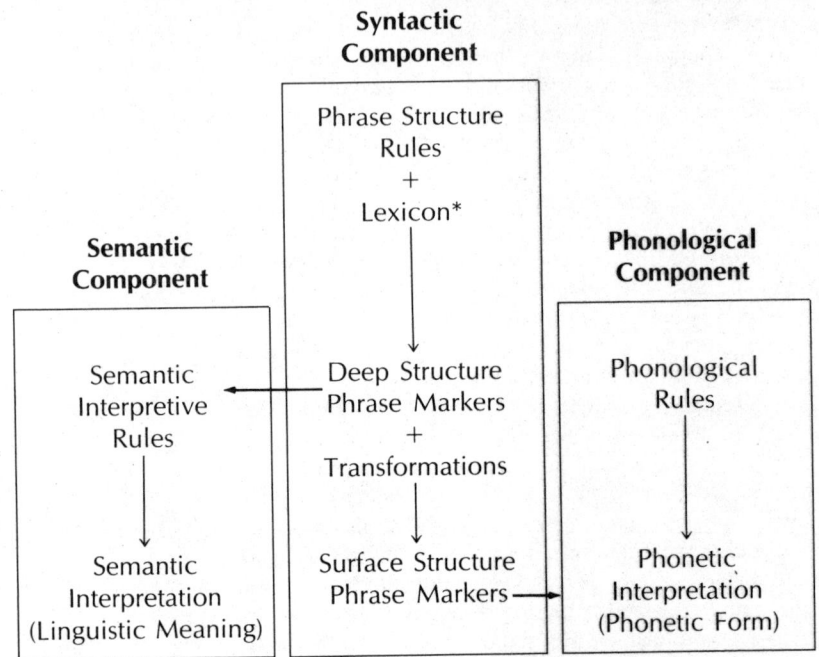

*A lexicon, a kind of internalized dictionary, is an unordered list of linguistic forms—together with essential information on the meaning and pronunciation—that are inserted into deep structure phrase markers underlying sentences; see the "Phonological Component" (pp. 1–6) and "Semantic Representation" (pp. 61ff.)

FIGURE 2. A Transformational-Generative Grammar (Standard Theory). A Model of Linguistic Competence

To make this discussion more concrete, we will sketch the two derivations from deep to surface structure for (5); let's say one will be (5′) and the other (5″). So (5′) with its paraphrase (6) will be derived in Figure 3, and (5″) and its paraphrase (7) will be derived in Figure 4. Then by comparing these derivations, as shown in Figure 5, we can see how the grammar accounts for the ambiguity of sentence (5).*

---

*Remember that (5′) and (5″), each with a distinct meaning, have the same form—a single string of words.

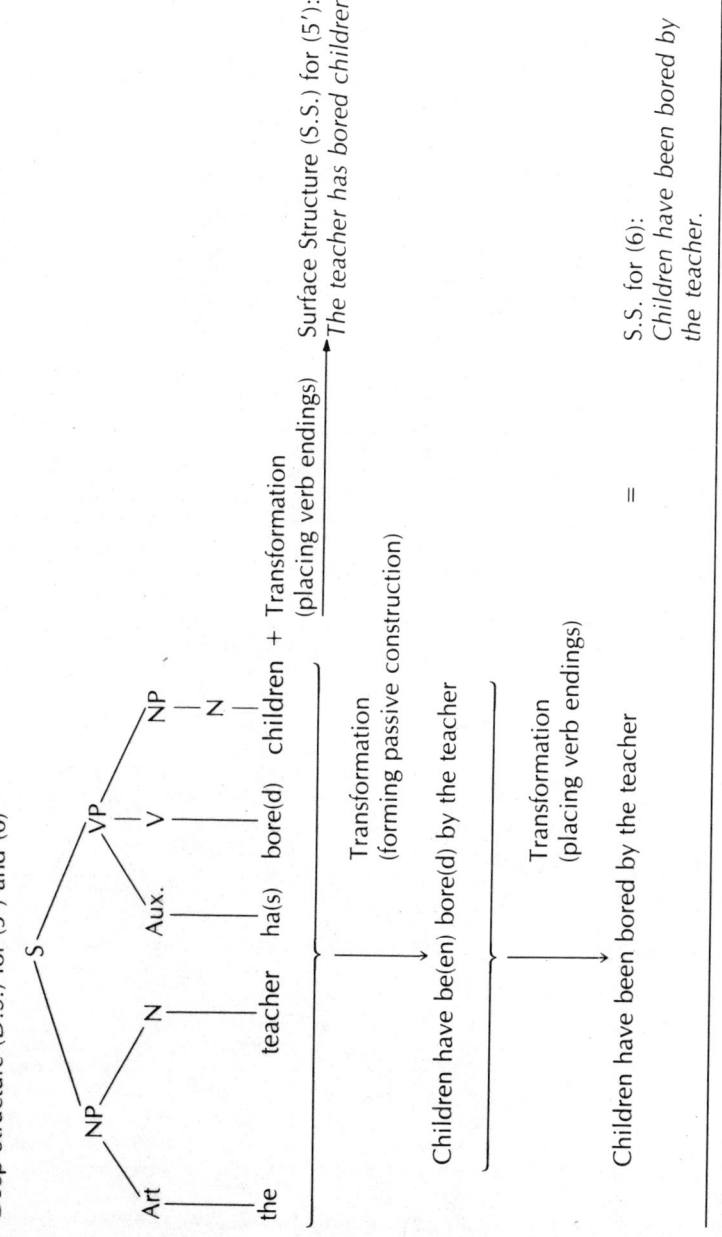

FIGURE 3. *Derivations for (5′) and (6) Paraphrase Set*

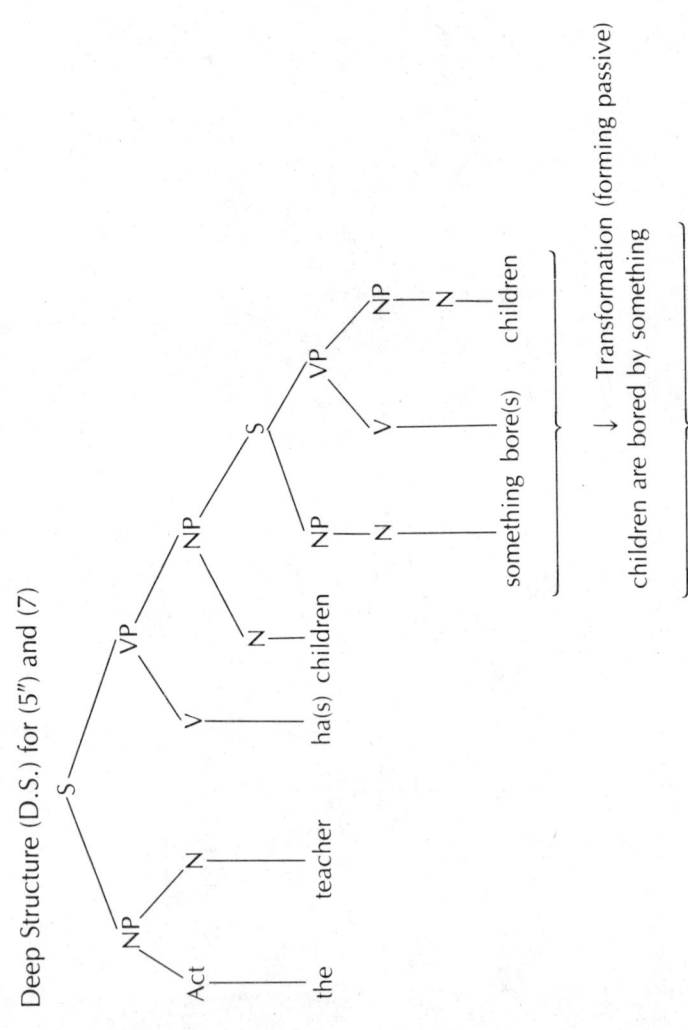

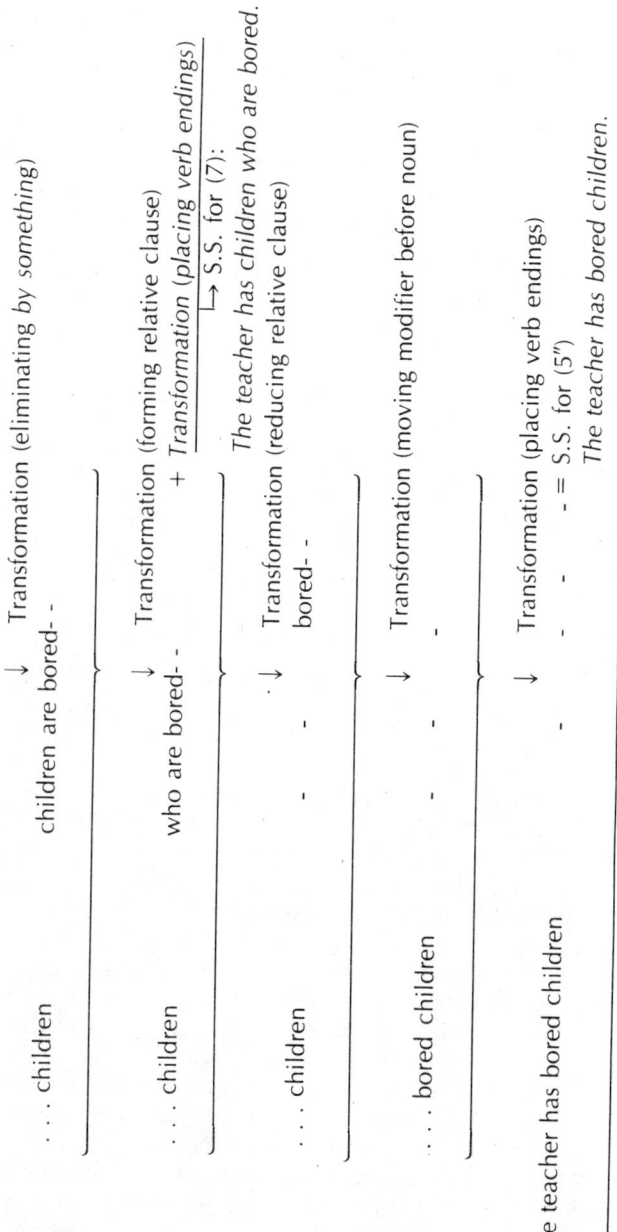

FIGURE 4. Derivations for (5") and (7) Paraphrase Set

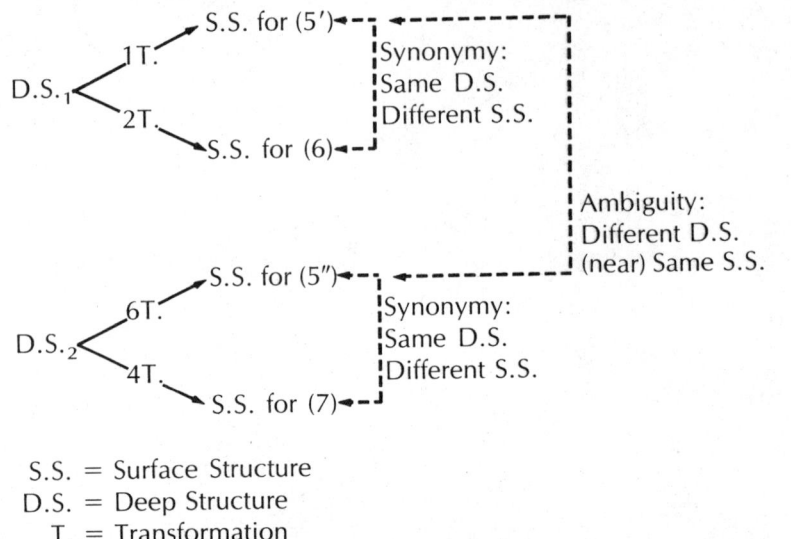

FIGURE 5. *Summary of How a Transformational-Generative Grammar Accounts for Ambiguity and Synonymy (Paraphrase)*

Without actually formalizing the phrase structure and transformational rules, we have provided a fairly detailed account of how the grammar represents our knowledge of sentences (5)–(7) in isolation, illustrating the ambiguity and synonymy relations. Yet even without an introduction to the formalism of the rules, the reader should come to a greater appreciation of the complexity and abstractness of the system of competence—the mental structures and abstract computations that figure into the determination of sentences in a language.[112]

The system of competence is seen to be all the more complex when linguists conclude that the set of transformations required to derive grammatical surface structures must be at least partially ordered to yield results that match native speaker judgments. Much of the research of linguists involves the determination of not only the rules themselves but also the order in which they must apply. However this knowledge is ultimately represented in the mind, the rule order requirement of the linguist's model is sufficient evidence to demonstrate that linguistic competence must be highly constrained knowledge, that is, a complex and rich system, in many respects still elusive of precise description. Space does not permit us to delve into the rule-order question concerning transformations; however, the reader is encouraged to investigate any of the following works on syntactic structure in English: Akmajian and Heny,[113] Culicover,[114] and Baker.[115]

To summarize, for any sentence we subconsciously assign an abstract deep structure (which determines the linguistic meaning) and a surface structure (which determines how the sentence may be pronounced); these mental structures* are related through abstract operations that grammatical transformations represent.[116]

The Phonological Component and Phonetic Representations

In chapter 3 (pages 49–50) we mentioned that our perception of sounds is dependent on the rules of our internalized grammar; among other things, we "pay attention" to certain phonetic features and "ignore" others. To cite another example, in the words *pin* and *spin*, the *p* segments are phonetically different, yet perceived as the same. Both sounds are produced with the air stream momentarily *stopped* by the closure of the lips; both are *voiceless* sounds, that is, without the vocal cords vibrating. Yet the first is accompanied by a puff of air, or *aspiration*, whereas the second is not.† This difference, though, is predictable: the aspirated segment is in initial position in the word; the unaspirated segment is noninitial. Thus, the phonetic feature of aspiration in voiceless stop consonants in English (*p, t,* and *k*) is predictable by a rule in the phonological component of the grammar.

More generally, we may say that phonological

---

*Some linguists (e.g., Fodor, Bever, and Garrett) attribute psychological reality to the structures themselves, but not to the rules.
†The reader may observe this phenomenon by holding a hand in front of the lips and saying both words.

rules add predictable phonetic detail to the segments in the surface structure. The lexical entries in a phrase marker are thus specified with only the nonpredictable phonetic features, that is, those *distinctive* features that may contrast one segment from another.* Accordingly, the distinctive feature bundle *bilabial, stop, voiceless* would characterize the *phoneme* /p/, whereas *alveolar, stop, voiceless* would represent the phoneme /t/, etc. The predictable or *nondistinctive* feature of *aspiration* would be added by a phonological rule that takes account of the position of the segment. The lexical items in the surface structure would be spelled phonologically, i.e., with distinctive-feature notation identifying the phonemes. Then phonological rules would apply to the structured string and add the nondistinctive features and other predictable information including the assignment of stress, ultimately yielding a full *phonetic* representation.

With respect to the theory of competence, the phonetic features may be considered sets of instructions to the vocal tract.[117] Linguists find that there are many general principles (statable as phonological rules) which derive the fuller phonetic representations from the more abstract phonological representations. These rules, together with the distinctive feature sets comprising phonemes constitute our knowledge of the sound system that serves as the basis of our production and perception of speech signals.

This discussion has necessarily been brief and oversimplified; the reader is therefore encouraged to look at any of the introductions on the subjects of the phonology and morphology of English by Langacker,[118] Akmajian et al.,[119] Fromkin and Rod-

---

*More properly, the lexical items are comprised of meaning-bearing units called *morphemes: roots* and *affixes* (*prefixes* and *suffixes*). For example, the *p* segment is distinguished from *b* by the voicing feature; the former is voiceless, while the latter is voiced; on the other hand, a *p* is distinguished from a *t*, in that the latter stop is made with the tip of the tongue against the gum ridge, not with the lips. For the sake of illustration, we may designate these distinctive features as *bilabial* (for *p*) and *alveolar* (for *t*).

man,[120] and Sloat et al.,[121] for a condensation of Chomsky and Halle.[122]

A Model of Linguistic Competence

Semantic Representation

Perhaps no other area of language has been argued over more intensely than semantics. The reason seems clear: meaning is the most intangible aspect of language. Traditionally, philosophers, logicians, and, more recently, social scientists have proposed accounts of meaning—theories of sense and reference of nameable things, of the truth value of sentences in isolation and in context, of the relation between an assertion and its presuppositions, etc. In the last quarter century American linguists have increasingly turned their attention to the description of the semantic content of utterances, considering the boundary between syntax and semantics and the difference between linguistic meaning and speaker meaning.*

There has been a great deal of debate among linguists concerning the scope of meaning that a grammar can account for—whether strict linguistic meaning can be distinguished from other aspects of meaning that may involve a speaker's knowledge about the world, beliefs, and *pragmatic competence*. The latter notion involves one's knowledge of appropriate usage in particular social contexts, intentions in speaking, and implications of what is said.

Concerning the relation between linguistic knowledge and knowledge of the world, we may ask whether the truth values of (1) and (2) below are determined on the same basis:

(1) The box that I lifted was too heavy for me to lift.

(2) The box weighed exactly 100 lbs.

Obviously, sentence (1) is a contradiction, i.e., lin-

---

*From the late 1960s to the late 1970s, many linguists proposed very abstract deep structures in order to account for increasingly larger conceptions of linguistic meaning. Certain of these proposals within "generative semantics" proved inadequate for lack of rule specification.[123] These questions are not necessarily unrelated, but they do underscore the complexity of the subject matter, as has been emphasized throughout this chapter.

guistically false by virtue of those lexical items occurring in that grammatical structure; on the other hand, the truth or falsehood of (2) would have to be determined on extralinguistic, or empirical, grounds.

With respect to pragmatic aspects of meaning, we observe that a sentence like (3) has a literal meaning, as determined by its grammatical structure, but it may also have a nonliteral meaning if we consider the speaker's intentions in a given context (in the performance of a *speech act*):

(3) You are wearing a lovely tie!

The speaker may intend the opposite meaning from its literal one, i.e., sarcasm.

Similarly, a speaker may intend for a statement like (4) to persuade the listener to perform an act that would more directly be requested by sentence (5):

(4) It's hot in this room.
(5) Would you please open the window?

Consider, finally, the culturally influenced implication (based on a belief system) of the following statement, directed toward an elderly individual:

(6) You are very old!

In many non-Western societies, this would be intended as a compliment; hardly so in the United States!

Because of the complex interplay between systems of belief, knowledge of the world, and pragmatic factors, along with one's linguistic competence, Chomsky, Jackendoff, and others[124] have argued that it is not possible to offer a linguistic account for all aspects of meaning. The thrust of the distinction is to isolate that part of the total meaning of an utterance in its context which is intrinsic to the knowledge of the language itself, what Chomsky calls "logical form."[125]

In the standard model of linguistic competence, deep structure determines the linguistic meaning of an utterance, that is, the meaning of the lexical items together with their grammatical relations in a deep-structure phrase marker. Therefore, we can say that the linguistic meaning of a sentence is *compositionally* determined—from the constituents in their constructions at an abstract level.[126]

The effect of the structural configurations at this level is crucial to semantic interpretation, as we have already demonstrated with the examples of the complexity of the sound-meaning relation in chapter 3; deep-structure grammatical relations define the *thematic* relations in the following manner: Taking the two thematic relations of *agent* and *patient* of the verb, as mentioned in chapter 3, we observe that sentences (7) and (8), while quite different in form, are paraphrases:

(7) The dog bit the mailman. (active voice)
(8) The mailman was bitten by the dog. (passive voice)

These sentences have different grammatical relations in the surface structure: *the dog* is the subject of sentence (7) but the object of the preposition in (8); *the mailman* is the object of the verb in (7) and the subject of sentence (8). Nevertheless, the thematic relations are the same: *the dog* is the agent of the verb *bite* and the *mailman* is the patient of the verb in both cases. Thus, underlying the two sentences, in the deep structure *the dog* will be the subject and *the mailman* will be the direct object, as sketched in Figure 6 (below):

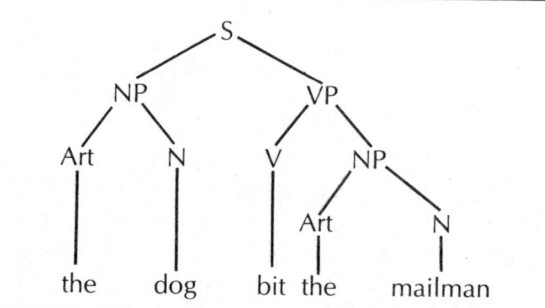

FIGURE 6. Example of Deep Structure

In the deep structure, then, there will be a correlation between the grammatical relations (subject, object, etc.) and the thematic or logical relations (agent, patient, etc.). The deep-structure subject will be the agent of the verb (that is, if agency is expressed), while the deep-structure direct object will be the patient of the verb. In order to derive the surface structure for sentence (8), a passive transformation will apply to the deep structure, altering its form by switching the two noun phrases, inserting *by* to make a prepositional phrase, and inserting a passive auxiliary with *be* and the *-en* past participle inflectional suffix.

The lexical items themselves are actually composed of smaller meaning-bearing units called *morphemes*, including *roots* and *affixes*. Roots may be *free* (words themselves), like *boy, egg,* etc., or *bound*, like *vis-* (from Latin, related to "sight"), as in *vision, visual*, thus requiring some affix(es) to form a freely occurring word. Affixes may be classified according to position with respect to the root as *prefix* or *suffix* (*infix* in some languages) and according to function, as *inflectional* (grammatical endings such as noun plurals and verb tenses) and *derivational* (forming different lexical categories like *-ize* for verbs, and *-ion* for nouns). There are only a few inflectional affixes in English, whereas there are numerous derivational affixes. The latter are attached to roots and larger stems in accordance with *lexical formation rules*.[127] Such rules account for our knowledge of the structure and meaning of lexical items that we have not necessarily "learned." For instance, given a verb of the form "*stem* +ize," we would intuitively know that there is likely to be a noun of the form "(——ize)+ion": "specialize"/"specialization."

To further illustrate this distinction between strict linguistic meaning and other (including pragmatic) aspects of meaning, we will analyze the following sentence in a little more detail:

(9) The actor regretted losing his cool.
   (i) Linguistic aspects of meaning:
      (a) Lexical meanings: Linguists like Fodor and Katz[128] have approached the sense of lexical items in terms of semantic feature (componential) analysis; thus, *actor:* male, human (the suffix *-or* would independently be identified as "one who + verb")

lose one's cool: (idiom, *cool* with metaphorical sense): be angry, not have control*

(b) Syntactic contributions:
   (i') Reference:
      the: Use implies speaker's assumption of listener's identification of the individual. (Equivalently, this could be considered an entailment (presupposition) relation under [c].)
      his: co-referential with the noun phrase *the actor* (only) PAST (tense): at a (remote) time before present
   (ii') Thematic relations:
      the actor: experiencer of verb *regret;*
      agent of the verb *lose*
      In the embedded sentence construction (*the actor's losing his cool*), the identical subject is eliminated; thus the experiencer and agent are identical individuals.

(c) Entailment (presupposition) relations:
   regret: Implies the truth of the embedded sentence because *regret* takes a "factive" complement.[131]

(d) Truth conditions: Following a *correspondence* theoretical approach, the statement (proposition) is true if it corresponds with a given state of affairs existing at a given time and place.

(e) Focus: Dependent on the location of the primary sentence stress, usually the last primary syllable in the utterance; i.e., *cool* in this case.

---

*These are rough approximations; see Kempson,[129] and Lyons,[130] on criticisms of this approach to semantic-primitive analysis.

(ii) Pragmatic aspects of meaning: ("speech act" features)
  (a) General:
    (i′) Speaker's intention: The speaker could use this sentence: to praise the actor for his rectification of aberrant behavior; to warn the listener of the actor's behavior, etc.
    (ii′) Assumptions for Successful Communication
    Speaker and Hearer make assumptions concerning (a) the truthfulness, (b) relevance, (c) appropriateness of the message.[132]
  (b) Culture specific:
    (i′) Conversational implication expressed in the idiom indicating that in (North) American/European societies, the value that one should not be emotional in making decisions; "coolness" of temperament falls within a certain norm of expectations, given the level and magnitude of the decision; hence, it is appropriate to regret such "aberrant" behavior.
    (ii′) Stylistic code: The use of the idiom implies an informality of style that the speaker uses in light of his or her assumption of what is appropriate for the listener.

This account is by no means exhaustive; it is intended to open up for the reader some of the complexities concerning semantic description of utterances in context. Kempson[133] discusses some other facets of meaning based on a totally tacit "pragmatic universe of discourse" that involves what the speaker believes the hearer believes (the speaker believes, etc.); such knowledge is part of the systems of belief and knowledge about the world (including the participants) that a full linguistic theory of performance must take account of. (For further study the reader is directed to the following works: Lyons, Kempson, Palmer, and J. D. Fodor.)[134]

We have surveyed certain features of human language that underscore the general observation that a native speaker's subconscious knowledge of grammar—not to mention other interrelated cognitive systems—is complex beyond anyone's full appreciation. A linguist's representation or model of this competence is therefore highly abstract, incorporating different kinds of rules and structures, thereby explicating our knowledge of form and meaning in sentences.

chapter five

Language Acquisition: Biological Background and Developmental Stages

The biological basis of language acquisition is discussed as well as the stages of language development in children from about one year through age six and on into puberty. Some strategies involved in such language acquisition are described.

The Language Gap

Language Complexity

We have seen that a *human* language is a highly complex, abstract system of knowledge involving the relation between sound and meaning. Even if we examine a single sentence, the set of rules required to account for the sound-meaning correspondence is of an impressive order. At every level, we observe language to be rule-governed.

Nevertheless, as we have also seen, this knowledge is not the exclusive property of a few exceptional individuals; rather, unless brain damage is involved, *all* children acquire a fully developed system of competence. In fact, as Eric Lenneberg* states, "a total absence of language development is nowadays seen only in the worst cases of feeblemindedness and chronic psychosis."[135]

It is therefore quite proper to ask at this point, How does this complex system of knowledge develop in the child? This question entails both biological and psychological considerations. Accordingly, we may inquire into the nature of the brain and language, into the developmental stages of acquisition, and, ultimately, into the possible explanations for language acquisition. In this chapter we will be concerned with certain biological aspects (as part of neurolinguistics) and with observed stages of development (as part of psycholinguistics). In the next chapter, we will discuss alternative hypotheses concerning the nature of the task of the child in acquiring language.

Some Biological Aspects of Language

Traditionally, there has been a long-standing question whether language is a general function of the human brain or whether only certain parts of the brain are involved. Correspondingly, there has been the question of whether language is a function of general intelligence or a specific faculty of mind. We turn to the latter question in the next chapter.

In 1861, French physician Paul Broca pointed out the relationship of a frontal area in the left hemisphere

---

*The late Harvard and Cornell physiologist who specialized in the biological basis of language development.

of the brain with some forms of language impairment, later termed "aphasia." In 1865 Broca reported that damage to areas of the left hemisphere would be followed by certain disorders in speech production, including slow and laborious articulation—"motor aphasia," as it would more broadly be called later. However, Broca also noticed that corresponding damage to the right hemisphere did not have these effects.[136]

Neurolinguistic research since Broca's time has substantiated the claim that language is left-hemisphere dominated. Harvard neurologist Norman Geschwind observes that 97 percent of permanent aphasia cases are caused by lesions in the left hemisphere.[137] Most of the right-sided lesions causing aphasia occur in left-handed individuals.[138]

"Broca's area" is adjacent to the motor cortex, which controls the speech muscles. It is not surprising, then, that damage to this area affects speech articulation. However, as Geschwind points out, damage to the motor cortex area itself does not result in permanent disability because, if necessary, the speech muscles can be controlled by the right hemisphere.[139]

Neurolinguistic research was furthered in 1874 by the German physician Carl Wernicke in his discovery that a lesion in another area of the left hemisphere could affect both speech comprehension (instead of production) and semantic intelligibility (instead of phonetic intelligibility). "Wernicke's area" is adjacent to the cortical region receiving auditory stimuli. Thus Broca's aphasia may be characterized by halting, sometimes "telegraphic" speech, whereas Wernicke's aphasia typically results in seemingly well-formed structural usage but with semantic anomaly— empty words like "thing" repeatedly used, roundabout expressions for exact terms, substitution of inappropriate words, etc.[140] Wernicke hypothesized, correctly, that Broca's and Wernicke's areas are connected by a bundle of nerve fibers, thereby establishing a relay system between the areas responsible for the production and reception of the auditory signals of speech. Geschwind provides a simple model that may help to show the relationship of the two areas:

Language Acquisition: Biological Background and Developmental Stages

When a word is heard, the output from the primary auditory area of the cortex is received by Wernicke's area. If the word is to be spoken, the pattern is transmitted from Wernicke's area to Broca's area, where the articulatory form is aroused and passed on to the motor area that controls the movement of the muscles of speech.[141]

One of the leading authorities on brain structure, British physiologist John C. Eccles* notes that although sensory and motor functions are located equally in both hemispheres, the asymmetry (hemispheric imbalance) with respect to language is surprisingly missing in nonhuman primates.[142] This left-hemisphere dominance in humans is physiologically evidenced by an apparent enlargement of the language area in the left hemisphere. Eccles points out, in fact, that human infants are born with this asymmetry: "It has already been grown by genetic instructions in anticipation of its eventual usage in speech, and can be detected even in a 5-month-old fetus.[143]

Further confirmation of the left-hemisphere dominance comes from split-brain studies and dichotic listening tests. The former research[144] concerns surgical disconnection of the two hemispheres—severance of the great band of nerve fibers connecting the two sides of the brain—thereby preventing the two hemispheres from "communicating" with each other. Since information from the left side of the body is received by the right hemisphere primarily, and vice-versa,[145] in such cases subjects could not talk about information from the left side, e.g., about an object in the left hand, because the information could not be transmitted from the right hemisphere to the left in order for the subject to verbalize about it.

Dichotic listening studies reveal that there is a right-ear advantage with respect to speech perception, since information from the right side is received by the left hemisphere.[146]

While the physiological asymmetry may be de-

---

*For more on Dr. Eccles's brain experiments, see Arthur Custance, *The Mysterious Matter of Mind* (Grand Rapids: Zondervan/Probe, 1980).

tected in a newborn, a process of lateralization (one-sided specialization) culminates at puberty.[147] This period has been described as the "critical period" for the spontaneous acquisition of the first language,[148] with most of it taking place by age five or six. A chief proponent of the critical-age hypothesis, Lenneberg based his conclusions on studies concerning recovery from aphasia in different age groups.[149] He wrote that if aphasia strikes a young child during or immediately after the age at which the native language is acquired, what had been acquired up to that point might be totally lost. However, soon the child will start over, repeating the stages of infant vocalization, perhaps at a slightly faster pace, until, according to Lenneberg, he has mastered the language. The process will be slower if the injured child is slightly older, between the ages of three and four, but if the injury occurs by the time of puberty, it is likely that the aphasic symptoms resulting from the injury will not entirely disappear.*

Lenneberg hypothesized that there is a period in infancy when the hemispheres have equipotential for language function.[152] He cited the data of L. S. Basser,[153] which indicated no great difference in the onset of speech between young children who suffered lesions in the left and right hemispheres during the first two years of life, before left hemispheric dominance is established. Thus "a lesion in the left hemisphere is apparently sufficient cause to confine the language function to the right side."[154]

---

*There has been some question, however, as to whether the complete lateralization of the brain around puberty constitutes a true biological constraint on native language acquisition, since second-language learning theorist Stephen Krashen (1973) interprets Lenneberg's data as supporting a critical age terminating at five.[150]

In contrast to native-language acquisition, second-language acquisition after puberty has generally been found to be more painstaking, often incomplete, depending on the age of the learner, the circumstances of learning, the motivation, etc. In the last decade, second-language learning studies have focused on the nature of the errors made by the learner, as well as in-between states of emerging competence in the second language ("interlanguage"), in an attempt to compare the two acquisition processes.[151]

However, Mildred Freburg-Berry, writing around the same time as Lenneberg (in the late 1960s), concluded that "the potential of the minor hemisphere for language learning is far below the pre-traumatic potential of the major hemisphere," i.e., the ability of the left hemisphere before the aphasia has begun. This was because, in her experience, children who have incurred serious and presumably permanent lesions in one hemisphere are still impaired in all facets of language, even up to the age of ten or twelve.[155] In fact, recent neurological evidence seems to indicate that total compensatory switching of language (from left to right hemisphere) is not possible: there are at least certain syntactic capacities that are not accessible to the right hemisphere, nor, apparently, is some semantic information resultant from these syntactic structures, even when the left hemisphere is surgically removed in infancy.[156]

The age of onset of various stages of language acquisition coincides with the development of motor skills and the period of flexibility in the development of the human brain system. Lenneberg describes these correlating stages of motor development, pointing out that such milestones are synchronic with, but not causal to, various stages of language acquisition.[157] For example, when a child is able to sit up and bear his weight when put in a standing position at about six months, he is able to babble one-syllable utterances, consonant + vowel (CV) types. When at about ten months the child is able to creep and take steps by holding onto someone or something, he is able to make sound play and distinguish certain words heard. By about twelve months, when the child is able to walk with/without help, he is able to utter the first words (reduplicated CV combinations) and understand both words and simple commands. By about eighteen months, when the child is able to grasp objects and walk with a stiff gait, he has an active vocabulary of up to fifty words, mixed with babbling of polysyllables that have a characteristic intonation pattern (of rising-falling and rising pitch contours, typical of statements and simple questions in English, respectively).* At this stage, the child is at the onset of two-word utterances marking the beginning of the acquisition of syntax, which will be described in the psycholinguistic literature reviewed in the next section.

With these and more detailed correlations between motor development and language acquisition, Lenneberg stresses the essentially *noncausal relation* between the two phenomena:

> The temporal interlocking of speech milestones and motor milestones is not a logical necessity. There are reasons to believe that the onset of language is not simply the consequence of motor control. The development of language is quite independent of articulatory skills (Lenneberg, 1962); and the perfection of articulation cannot be predicted simply on the basis of general motor development. There are certain indications for the existence of a peculiar, language-specific maturational schedule. . . . Speech, which requires infinitely precise and swift movements of tongue and lips, all well-coordinated with laryngeal and respiratory motor systems, is all but fully developed when most other mechanical skills are far below their levels of future accomplishment.[159]

Lenneberg's last point is dramatically borne out by the fact that the same three-year-old child who employs hundreds of intricate muscle movements in making an utterance will also spill a glass of milk more often than not!

Having covered certain essential aspects of the language and brain relationship, and having shown that there are certain maturational coordinates (motor-skill correlates), we now turn to a more detailed discussion of the stages of language acquisition in children that have been identified by psycholinguistic research over the past twenty years. (The reader may wish to look further into neurolinguistic studies by considering the following general accounts of major trends: Akmajian et al., 1979,[160] Fromkin and Rodman,

---

*Taken from Lenneberg's data in *Biological Foundations of Language*, table 4.1, pp. 128–30. It should be stressed that these dates given by Lenneberg are approximate.[158]

1978,[161] in addition to the other references in this section.)

Developmental Stages in Child Language Acquisition

Recalling from chapter 3 that by about age six a child has acquired most of the rules of an adult grammar, we now proceed to review some of the psycholinguistic research of the last twenty years that deals with the accomplishment of such a monumental task. Largely because of the impetus provided by transformational linguistic theory concerning competence, there have been numerous cross-sectional and longitudinal studies, including some with children learning non-Western languages.[162] These studies have attempted to determine the following: the developmental stages, the psychological reality of the rules assumed to be acquired by the child, the strategies employed by the child, the relation of language acquisition to other cognitive development, and finally, the principled basis (explanation) for the accomplishment of language acquisition (the subject of the next chapter).

Although these issues are still under investigation, there has been a general consensus concerning the overall pattern of language development—that it proceeds in quite distinct and predictable stages, rather than gradually or haphazardly.[163] This we would expect; if, as linguists say, an adult native speaker's knowledge consists of rules, then a child will acquire the language in stages—in accordance with the acquisition of rules by applying and later revising them, until ultimately the emerging grammar resembles an adult's.

Developmental psycholinguistic research has concentrated heavily on the language acquisition of children from ages one to five, the major segment of the critical age. A few studies, such as Carol Chomsky's (1969) study of children's syntactic development from ages five to ten, have dealt with certain final phases of first-language acquisition—the "fine tuning" of an elaborate code.[164]

In the earlier work of the 1960s, a small number of developmental stages were plotted according to the syntactic complexity of children's utterances. Such stages were seen to correlate with chronological age fairly closely, but in recent years researchers have found more variation than was previously thought to be the case. Thus many have followed psycholinguist Roger Brown[165] in using another index to characterize the stages: Mean Length of Utterance (MLU)—measured in terms of the number of morphemes* per utterance—based on the assumptions that children's utterances get progressively longer and that length is a function of complexity as well.[166] Accordingly, within a particular stage (MLU Range) children tend to exhibit characteristic syntactic, phonological, and semantic features of the language, as exemplified in Brown's study of the acquisition of certain grammatical morphemes[167] in the earlier stages, as will be discussed under "Beyond Telegraphic Speech," pages 84–87.

However, for the sake of convenience, we will refer to general stage designations and approximate age ranges. These stages, based essentially on speech production, include the following: the babbling stage (generally considered the prelanguage stage), the one-word ("holophrastic") stage; the two-word stage (early syntactic stage); the so-called "telegraphic" stage; a middle syntactic stage; a later syntactic stage, approaching adult-like competence; and, finally, essentially full competence.[168]

Babbling Stage

Infants, including deaf babies, begin vocal expression by cooing for the first few months. Then, around the sixth month, a period of babbling is observed as a prelanguage stage, in which children seem to experiment with all kinds of sounds, and, in one view, suppress the sound patterns not used in their linguistic environment.[169] During this period, we observe that the intonation patterns (pitch contours) of adult speech begin to emerge. In English, this means that a rising intonation (used in simple questions like

---

*Recall that a morpheme is a minimal meaning-bearing unit of form, such as *the, write,* or the *-ed* of *waited.*

"*Is he busy?*") and rising-falling intonation (used in declarative statements like "*He's busy.*") would begin to be discerned in the child's sound play.

One-Word ("Holophrastic") Stage

After a few months (around the twelfth month), an interesting phenomenon takes place: among the nonsense syllables the child utters, faint likenesses to an actual word may be discerned, typically as he attempts to identify some familiar object or individual in the environment. The phonetic forms of such words may or may not be close to the adult version, but that the child has so spoken is not seriously in doubt (at least to the parent!).

Early Semantic Development. When the child begins to form his first words, he refers to some nameable entity, based on a concept of enduring objects. This remarkable ability to name something involves the phenomenon of *overgeneralization:* "da (-da)" may signify, not only the child's father but also other individuals (males) for some time. Psycholinguist Eve Clark proposes that a child first learns the *general* semantic (lexical) features of a word, e.g., *adult, male,* for "da-da," thereby overextending the usage of the word. Later the child acquires more *specific* semantic features, as found in an adult's internalized lexical representation of the word.[170] In the process, then, the child narrows the range of reference to correspond to the adult's. Another psycholinguist, Susan Carey, has revised Clark's missing-feature theory[171] to offer an account of the tremendous growth of vocabulary in child language.*

How, then, does a child begin to acquire these general semantic features that are successively refined? Clark,[173] following German linguist Manfred

---

*Questions regarding the discovery of semantic features in the first place, their relation to the conceptual (including perceptual) system, and the organization of lexical entries (words) given the semantic features, have concerned much recent discussion on child language development as well as more traditionally philosophical debate on what is nameable in principle.[172]

Bierwisch,[174] suggests that there is a universal set of semantic primitives (basic concepts) available to the child, as beginning points for the child to discover the general features of his first words. Beyond the initial naming process, children appear to use their one-word utterances to communicate some message; e.g., "milk" may mean "I want milk," or "I see milk," etc. Since these one-word utterances seem to stand for sentences,[175] this stage has often been called the "holophrastic" stage. How much children intend by their one-word utterances vs. how much meaning an adult imputes to them has been more openly discussed.[176] Nevertheless, it must be recognized that children at this stage of production understand a great deal more than what their utterances would indicate.

Early Phonological Development. In these monosyllabic utterances, we can see the beginnings of the acquisition of the phonological system of the language. The distinctive feature system, discussed in chapter 4, has been employed by its principal architect, the noted linguist Roman Jakobson, in describing a child's phonological development.[177] Essentially, Jakobson states that a child's acquisition of the sound system of any language is contingent on the mastery of the distinctive features pertinent to that language, a subset of a universal set. Moreover, he makes the strong claim that the stages of acquisition are predictable, based on the maximal phonetic feature contrasts—major distinctions are acquired before minor ones. Thus the elementary (CV) syllable /pa/ represents "polar configurations in the vocal tract"— the complete closure of the oral tract in the consonant versus the complete opening in the vowel.[178] Successive feature contrasts ("oppositions") are gained among consonants like oral versus nasal, e.g., /p/ versus /m/; as well as the opposition of labial versus dental consonants, e.g., /p/ versus /t/; peripherally articulated /p/ and /k/, front and back stops respectively, versus medially articulated /t/, etc. Moreover, similar differentiation among vowels can be seen, e.g., open /a/ versus closed /u/ and /i/ (discussed

79

Language Acquisition: Biological Background and Developmental Stages more technically in Jakobson and Halle, *Fundamentals of Language*, pp. 37–40).

The phonemes /a/ - /i/ - /u/, exhibiting the major distinctive feature contrasts for vowels, would therefore be predicted to be acquired before other vowel segments (which would be in between). While some researchers have found counterexamples to Jakobson's timetable of acquired features,[179] the theory, based as it is on distinctive feature notation used in generative phonology, provides an interesting hypothesis concerning the acquisition of sound features in a child's emerging competence. If this theory is true, then it can be said that a child maintains the sounds of the language he is exposed to and suppresses the others (available from the universal set), as mentioned in the introductory paragraph of this section (page 76).

Jakobson's theory of distinctive features does provide insight into a child's acquisition of phonemic contrasts, e.g., /b/ vs. /p/, where the voicing feature is distinctive; once a child acquires the above phonemic contrast, the distinction between voiced-voiceless pairs suddenly appears, e.g., /t/ - /d/; /k/ - /g/.[180]

A child's ability to articulate phonemic contrasts in language lags behind his perception of the differences. Thus a child may not be able to articulate the difference between, say, "dog" and "duck" (both vowel and final consonant contrast), but can nevertheless distinguish the difference in the adult's pronunciation of the two words.

Two-Word Stage: Early Syntactic Stage

Somewhere around the age of two, children begin to put two words together.* In the early 1960s, following the work of Martin Braine[182] on English phrase structure, psycholinguists began to describe

---

*This stage corresponds, approximately, to Brown's Early Stage I, where the M.L.U. range is between 1.0 and 1.5.[181] In fact, Brown includes this stage within "telegraphic" speech, described in the next section.

such two-word utterances as occurrences of a *pivot* word plus an *open* word. Thus Braine outlined a simple pivot-open grammar that children supposedly internalize.

Following are some examples of pivot and open words taken from studies summarized by psycholinguist David McNeill:[183]

Pivot	Open
allgone	boy
byebye	sock
big	boat
more	fan
pretty	milk
my	hot
see	mommy
this	.
the	.
.	.
.	.

FIGURE 7. Examples of Pivot/Open Words

Thus children at this stage are observed to form novel utterances by combining a pivot word and an open word, or two open words.

However, since the work of psycholinguists Lois Bloom and Melissa Bowerman in the early 1970s,[184] many child-language researchers have seen that pivot-open grammars are fundamentally inadequate to represent the competence of children at this stage. First, the analysis is based solely on patterns of occurrence in *surface* structure, without regard for the semantic relations that might indicate the different *deep structural arrangements*. Second, the pivot-open grammar itself is based solely on children's observed production without considering their ability to interpret more complex adult utterances. Thus the proposed grammar does not accurately reflect children's competence at this stage. In citing her own work with children learning Finnish (a non-Indo-European language) as well as studies on Samoan, Melissa Bowerman concludes that "the class of children's early two-word utterances were found to be much more complex than the two-class pivot-open model suggests."[185] What these psycholinguists have argued for, therefore, is a method of "rich" interpretation of children's early utterances, attributing semantic relations (found in adult usage) to their two-word sentences.*

To illustrate, various semantic (thematic) relations may be observed from children's two-word utterances. Brown lists eight main ones,[186] four of which are illustrated in sentences (1) through (4) below, occurring in the speech of a child from the ages of twenty-two to twenty-four months:†

(1) Me go.   (Agent-Action)
(2) My book.   (Possession)
(3) Hot tea.   (Attribution)
(4) Uncle funny. (Prediction)

We may accordingly surmise that even at this early syntactic stage, a child's emerging competence is more complex than what the forms of the sentences would suggest. There are aspects of an intricate conceptual structure that are thus being mapped onto an emerging linguistic computational structure.‡

The linguistic creativity of a child even at this early stage must again be stressed. Sentences like "Allgone milk" reflect a different word order from what a parent would utter ("The milk is all gone"). Yet, sentences like "Allgone outside" (meaning that the child could no longer see anything outside after the

---

*This matter will again be raised in connection with an analysis of ape signing in chapter 8, pages 126–34.

†Unless otherwise indicated, the sentences given in the rest of the chapter illustrating the various stages are taken from data collected by one of the authors involving two children. Sentences (1)–(4) occurred in the speech of one of the children, from the ages of 22 to 24 months. Contexts for these utterances have not been provided.

‡This distinction, made by Chomsky (*Rules and Representations*, pp. 54–57), will be taken up in the following chapters.

door was shut) reflect an even more significant facet to a child's ability to produce novel sentences.[187]

"Telegraphic" Speech

Following the two-word stage, usually somewhere after two years of age,† the child begins to form longer utterances that reflect the essential word-order characteristics of the language, e.g., Subject–Verb–Object (S–V–O) in English. However, most grammatical morphemes—the inflectional suffixes like tense endings on verbs and plural and possessive noun endings in English, as well as *function* words like prepositions, articles, and auxiliary verbs—are not found in these sentences. Thus, the term *telegraphic* was applied in the 1960s to draw a parallel between the child's general form of such sentences and the deliberate omission of certain grammatical morphemes in telegraphic style.

Here are a few example sentences observed during this stage:*

(5) Timmy eat it.
(6) Go wash it.
(7) Where grandma shoes?
(8) I no go bye-bye.

Despite the absence of grammatical markers such as verb tense (noun plurality is marked in sentence [7]) and function words like prepositions, auxiliaries, articles, etc., one can see a remarkable similarity between these sample sentences with adult-type sentences. For example, sentence (5) exhibits the S–V–O word order of major constituents in English; sentence (6), an imperative, also reflects the S–V–O word order, with deleted subject; sentence (7) exhibits the fronted *wh*-question word, though without the auxiliary; finally, sentence (8) shows the placement of the negative particle *no* before the main verb.

---

†This stage corresponds, roughly, to Brown's Late Stage I, where the M.L.U. range is 1.5–2.0, and possibly Early Stage II, up to M.L.U. 2.5.
*These sentences occurred in the speech of the two children from 22 to 24 months of age.

Many researchers, such as Ursulla Bellugi,[188] observe children first negating a sentence by prefixing the sentence with *no*, as in (9):

(9) No Mom sharpen it.

Then, later, children place *no* before the main verb, as in (8); then they acquire *don't* and other contracted negative auxiliaries, as in (10):

(10) I can't catch you.

They do this whether or not they have already acquired the affirmative auxiliaries first.[189]

Beyond Telegraphic Speech: Middle Syntactic Development

Once a child embarks on the learning of syntactic structures, there soon appears the basic order of constituents, with various semantic notions attached to the subject and object functions. Then various grammatical suffixes and function words "fill in" the essential structure of affirmative statements and imperatives, along with modification in word order for negative statements, simple (yes-no) and *wh-* questions.*

We may suppose, therefore, that the child has acquired certain grammatical transformations, such as sentence negation (in the stages noted above), auxiliary-subject inversion for questions, etc.[191] To say that a child has acquired certain transformational rules does not necessarily mean that he or she goes through a corresponding series of mental operations in either producing or comprehending sentences, as was advocated by psycholinguists in the 1960s. These researchers held to the *derivational theory of complexity;* i.e., a sentence was thought to be more complex to process than another if there were more transformations involved in its derivation than those in the other.[192] Lack of empirical support for this hypoth-

---

*This period ranges over several of Brown's stages, II–V, involving in his case studies a period of up to a year or so.[190]

esis underscored the distinction by Chomsky between a theory of competence and a theory of performance in the sense that a linguist's model of linguistic competence is not meant to be a model of how a native speaker actually processes sentences.[193] Nevertheless, questions concerning the psychological reality of the internalized rules continue to arise in the debate among linguists† over how best to formulate the model of competence.[195]

Other evidence of rule acquisition can be seen in the phenomenon of *overgeneralization:* in acquiring a new rule children have been found to apply it beyond its scope in an adult's grammar. Thus children at a certain point may abandon a correct form such as an irregular past tense verb like *went,* and use *"goed"* for a time because the latter conforms to the newly acquired rule for regular past tense formation.[196]

In this stage beyond telegraphic speech, we find children acquiring various grammatical morphemes, including noun and verb suffixes, as well as function words such as prepositions, articles, and certain verb auxiliaries. In his longitudinal study of developmental stages of acquisition of three children, Brown (1973)[197] provides a detailed account of the first two of five stages, which are based on his Mean Length of Utterance (MLU) measure, as previously mentioned. In Stage II, roughly corresponding to the late telegraphic stage and beyond, Brown examines the acquisition of fourteen grammatical morphemes.

1. progressive *-ing* verb suffix.
2. preposition *on.*
3. preposition *in.*
4. plural *-s* noun suffix.
5. irregular past tense suffix.
6. possessive *'s* suffix.
7. (copula) verb *be* (not contracted).
8. articles *a* and *the.*

---

†How a child infers what the rules are from the signals of surface structure remains largely a mystery, though some psycholinguists have outlined some possible strategies of sentence perception and production.[194]

9. regular past tense -*ed* suffix.
10. third-singular present tense -*s*.
11. third-singular present tense irregular.
12. auxiliary *be* (not contracted).
13. copula be (contracted).
14. auxiliary be (contracted).

Brown reaches an interesting conclusion: "The developmental order of the fourteen morphemes is quite amazingly constant across these three unacquainted American children,"[198] that is, in essentially the order given above. As can be seen, among the first morphemes to be mastered were the -*ing* progressive suffix (without the auxiliary *be*), the prepositions *on* and *in*, and the plural noun suffix -*s*.[199]

Concerned about how this particular sequence of learning might be accounted for, Brown concludes that the order in which the children acquired these morphemes can be generally predicted from a notion of structural complexity (similar to the derivational theory), employing a particular version of a transformational-generative grammar.*

Finally, although there was a fairly wide range of rate of development among the children in his study, Brown observes that age plus M.L.U. is a better predictor of level development than M.L.U. alone.[203] Again, we can see that there are stages of language development that in fact are related to chronological stages, despite the variations observed in these detailed studies. Such a correlation, together with the early language-motor coordinates of Lenneberg's

---

*The notion of *cumulative complexity* states that "a construction *y* is more complex transformationally than a construction *x* only if *y* involves all the transformations involved in *x* plus one or more others.[200] In contrast to the earlier *derivational theory of complexity*,[201] Brown states that his concept "does not assume that transformations all add a constant increment of complexity."[202]

Brown mentions that the construct of semantic complexity might also be seen as a determinant, since essentially the same predictions concerning the order of morpheme acquisition would be made. Actually this question appears to concern the relation of cognitive development and linguistic encoding of certain semantic notions.

study, strongly suggests that human language acquisition proceeds according to a maturational timetable.

Later Syntactic Development: Approaching Adult Competence

Children's sentences increase in seemingly geometric proportions once coordinating and embedding processes are acquired. One type of sentence embedding is noun-phrase complementation, in which a sentence is embedded in a noun phrase, typically the object first as illustrated below in (11)–(15):*

(11) Mommy said I can have it.
(12) I know how to do it.
(13) I wanna see what's in it.
(14) I think I lost my license.
(15) I don't know where it is.

Sentential expansions also take place in adverbial phrases, as in (16)–(21):†

(16) Don't put it in because the necklace might break.
(17) You juggle like Uncle Ken did.
(18) You better have this so you can chew it.
(19) You can use green if you want.
(20) When I was a little girl, I got shots too.
(21) This one is bigger than the wall.

Moreover, in this later period approaching full grammatical competence, children are observed to show increasing mastery of the auxiliary system; note the following hypothetical and low probability concepts signaled by the modals *could, would,* and *might*:‡

(22) I wonder if you could get it.
(23) If I were an elephant, he wouldn't bite me.
(24) You might be able to sleep well and be healthy.

---

*These sentences occurred in the speech data of one of the two children; age: 25–29 months.
†From the same child, age: 26–31 months.
‡From the same child, age: 36 months.

Thus children continue to acquire rules that more accurately reflect adult competence, until, around age five or six, one cannot easily recognize major differences either in the constructions they form (syntactically and phonetically) or in their interpretations of adult sentences, except for a few (such as certain forms of passive and some opaque constructions involving the infinitive mentioned in chapter 3).[204]

The conceptual structures that are observed, as evidenced in the sentences above, have caused psychologists and psycholinguists to ponder the relation between language acquisition and overall cognitive development, as well as to consider alternative hypotheses concerning the nature of first language learning.

In this chapter we have seen that language is primarily a left hemisphere brain function, that the essential period of acquisition takes place well within the period of brain lateralization by puberty, and that the data from child-language research strongly indicate the acquisition of rules in discernible stages within a certain maturational schedule.

We are now ready to investigate the question concerning a principled explanation for this amazing development of abstract knowledge that takes place in only a few years. Does the child learn the whole system of competence from his environment based on some general learning strategies or is he innately endowed with some form of knowledge of language principles that makes language acquisition possible? Finally, it is noteworthy to mention that in the study of human language acquisition, we have come to the question concerning innateness. Considering the data on the ape communication experiments, summarized in chapter 2, we might ask if the same kind of question could be raised regarding innateness in animals.

chapter six

Theories Concerning Language Acquisition

Here the authors compare the major theories of language acquisition: their origins, presuppositions, and general predictions. The major issues concern the universality, complexity, and species-specificity of language acquisition and whether or not empiricist "blank slate" theories are adequate to account for these facts.

In approaching the various theories of language acquisition, it is important to reemphasize three facts: (1) the uniformity of language acquisition, (2) the complexity of the system acquired, and (3) the relatively short period of time involved in the major phase—five or so years.

The uniformity of language acquisition is reflected

in (1) the universality of the process—*all* normal children learn the language they are exposed to, barring those with severe brain damage causing feeblemindedness;* (2) the general stages each child goes through, also as discussed in chapter 5; and (3) the completeness of the linguistic task: all children acquire a full system of competence. We simply do not find people who lack any rules of the dialect they have been exposed to.

Concerning the complexity of the system acquired, it should be remembered that we have presented only *some* of the rules in English in chapter 5. All languages are so complex that there is still no complete grammar written of *any* language—even English, surely the most analyzed language of all. We pointed out in chapter 3 how linguistic competence is subconscious and thus how inaccessible the knowledge of a native speaker is. We will also show later in this chapter that there is an even greater degree of abstractness to language than what could be supposed from the foregoing analysis.

Finally, concerning the time factor, as mentioned in the previous chapter, the fact that certain children are faster or slower than others in acquiring certain features of the language is *not* a strong claim for significant variation among children, because only a relatively few years are involved anyway. Essentially, by the time children reach kindergarten, they have acquired a system of far greater magnitude than anything presented formally in school. Thus the difference of some months, or even a year or more, is not extremely significant in the final analysis.

In light of these observations, it is indeed striking to find the empirical conditions surrounding a child's acquisition of language to be so diverse: (1) differences in external conditions, (2) differences in general intelligence, aptitudes, etc., and (3) differences in the percentage of grammatical vs. ungrammatical/semigrammatical sentences (the samples of the language) that the child is exposed to, attesting to the

---

*See Lenneberg's statement on page 70.

somewhat "degenerate" nature of the primary data from which a child acquires language. Concerning factor (1), we observe that children are raised in varying home conditions, ranging from the absence of siblings (an only child) to the absence of parents (as in an orphanage). Besides, there are variations in emotional and personality factors among family members, attitudes toward learning, felt needs, etc. Parental input, if available, may also vary greatly, ranging from continual "correction" to near-neglect.

Concerning (2), the variation in general intelligence, we reemphasize that only children with some sort of brain damage or genetic deformity do not fully acquire a language; otherwise, regardless of level of intelligence (within the normal range), we find that *all* children acquire a full system of language competence. In contrast, we observe that there is great variation with respect to other types of knowledge acquired, say, knowledge of mathematics.

With respect to (3), the nature of the data, it is important to mention that children will hear samples of the language from various sources—peers; younger and older children; and parents, who speak to children and to other adults. Thus children hear a lot of "baby talk," which is somewhat deviant from fully grammatical usage, e.g., characterized by the absence of articles, morphological tense markings, etc. Additionally, both peers and adults make performance errors (see Language Structure Versus Language Use, pp. 48–49 in connection with linguistic performance). Thus children will hear both fully grammatical and semigrammatical sentences, but they will not necessarily be told what the difference is; i.e., adults do not usually signal their deviant utterances; certainly other children do not. The amount of grammatical deviance is obviously not quantifiable.[205] Nevertheless, children *will* acquire a basically *correct* grammar of their native language (particular dialect), regardless of the percentage of ungrammatical utterances they hear.

What this means is that children are not presented with an *ideally* suited array of data, as if one were to be *taught*. People may address them in baby talk, in sentence fragments, including elliptical (shortened) forms, and (unless addressed in some deliberately modified form) at normal conversational speed—the "stream of speech."

Moreover, in many parts of the world, including subcultures within the United States, there are children who, being exposed to more than one language or dialect, grow up bi- (or even multi-) lingual/dialectal.

Empiricist Theories of Language Acquisition

Ever since John Locke proposed that knowledge is gained from experience—through the senses—written on an essentially "blank state" of a mind, various notions of how learning takes place have dominated Western philosophies of knowledge (epistemology), known as *empiricism*. Modern empiricism has taken the form of *behavioristic* psychology, one school of which is led by B.F. Skinner, whose version of conditioning is applied to language learning and use in his book *Verbal Behavior*.*

On the assumption that a child brings to the acquisition of knowledge nothing unique, save some learning strategies, empiricist theories have emphasized, among other things, the role of imitation of parents' models, analogic creation ("stimulus generalization"), and reinforcement. We now turn to these three phenomena in particular.

Imitation

On first sight, imitation appears to be the main mechanism children employ in learning a language from others in the speech community. They are observed to mimic words or short utterances of parents or other (usually older) siblings. However, as has been pointed out by psycholinguists, imitation is screened through the emerging grammatical stage children are "working with," as is evidenced from

---

*Skinner's book *Verbal Behavior* (New York: Appelton-Century-Crofts, 1957) is a strict behaviorist account of verbal behavior and learning.

their early attempts to imitate the first forms of words; certain sound features are acquired before others; thus perfect imitations are rare in the early stages.[206]

Yet, this very point of the filtering of imitation is often overlooked by those who attribute so much language learning to this strategy. DeVilliers and deVilliers point out that imitation is essentially a *selective* process.[207] If we observe children in the early and later telegraphic stages (see One-Word Stage, pages 78-80), "not everything the child hears gets imitated."[208] What children *already* know (subconsciously) appears to be a major determinant of what they will imitate.[209] Additionally, children have been found to vary greatly in their spontaneous imitations, and yet what *is* imitated tends to be those features that have just begun to appear in free speech.[210]

That imitation cannot be an adequate account of a child's acquisition of a language is also evidenced by the obvious *novelty* in language use, even during the two-word stage. First, we notice that the word order in a child's utterances might differ from an adult's:

(1) (a) Child: Allgone milk.
    (b) Parent: Yes, the milk is allgone. (= an adult's *expansion* of a child's utterance)

We also notice that many such two-word utterances are indeed ingenious:

(2) Allgone sticky.[211]

Further, we note that the most frequently occurring words in adult speech—the function words like prepositions and articles—are *not* the first words used by children. As we saw in chapter 5, it is not until the telegraphic stage that children begin to use articles and some prepositions. Moreover, imitation fails to account for the fact that children often abandon correct forms (like irregular past-tense inflections) at points when they are acquiring a rule (like the regular past tense "-ed" suffix). This is the phenomenon of *overgeneralization* referred to previously.

Finally, Brown concludes that with respect to the fourteen morphemes used in his investigation,[212] "there is no clear evidence at all that parental frequencies influence the order of development of the forms we have studied. . . ."[213]

Analogic Creation ("Contextual Generalization")

Another notion of language learning involves the view that language is a set of patterns that can be acquired through analogic creation, or "contextual generalization."[214] The child perceives a pattern and generalizes to form an analogous sentence. Thus on hearing (3) below, the child might later say (4):

(3) Daddy go bye-bye.
(4) Doggie go bye-bye.

While this strategy implies a more sophisticated cognitive apparatus than what might be called for in the previous case (and such knowledge could be considered innate), it is still insufficient to account for the acquisition of a system of knowledge far more complex than the knowledge of a set of patterns observable in surface structure. Why doesn't a child generalize (8) from (6), analogous to (7) from (5)? In point of fact, as discussed in chapter 3, (5) and (6) have different deep structures, illustrating the opacity relation (see The Complexity of Language, pages 45-46):

(5) Ice cream is fun to eat.
(6) Ice cream is sweet to eat.
(7) It's fun to eat ice cream.
(8) It's sweet to eat ice cream.

Certainly, as formulated, this empiricist notion cannot account for the acquisition of an abstract level of deep structure (vs. surface structure).*

In short, we agree with Chomsky's conclusion, "To attribute the creative aspect of language use to 'analogy' or 'grammatical patterns' is to use these terms in a completely metaphorical way, with no clear sense and with no relation to the technical usage of linguistic theory."[216]

---

*A discussion of the inadequacies of contextual generalization is presented by Bever, Fodor, and Weksel.[215]

Reinforcement

In keeping with stimulus-response learning theories, the mechanism of selective reinforcement is put forth as an account of a child's language acquisition, similar to the conditioning of animals, as in Skinner's previously cited *Verbal Behavior*. By emphasizing the role of *external factors*, such studies minimize the contribution of the learner. Children's correct utterances are said to be positively reinforced/approved and thereby strengthened, whereas incorrect utterances are weakened with negative reinforcement (disapproval).

Psycholinguists have shown serious limitations with reinforcement learning theory: they have found that parents do not necessarily correct children's sequences because of syntactic or phonological deviance; rather, they might correct utterances according to the truth-value.[217] Instead of being guided by parental modeling, children seem to set the pace according to the rules they are acquiring and according to their maturational development. Certainly their acquisition in stages indicates that they are acquiring rules. In fact, overt attempts by parents to "correct" children grammatically have often been observed to be essentially futile.[218]

Chomsky's review of Skinner's book makes clear that the empiricist notions of *reinforcement, conditioning,* etc., really have no explanatory value: "The insights that have been achieved in the laboratories of the reinforcement theorist, though quite genuine, can be applied to complex human behavior only in the most gross and superficial way, and that speculative attempts to discuss linguistic behavior in these terms alone omit from consideration factors of fundamental importance...."[219]

Thus these and other empiricist theories fail to adequately account for the complexity and abstractness of the system of linguistic rules that children acquire. While there undoubtedly is a place for imitation, analogic creation, and reinforcement, these are clearly insufficient as theories of language acquisition. We now turn to the other alternative, which attributes

Rationalist or "Nativist" View ("Innateness Hypothesis")

innate structure to the child, emphasizing the contribution of the learner.

Recalling some of the facts of language learning reviewed in the discussion on the empirical theories of language acquisition above, we ask *why* there is such uniformity to language acquisition. Why don't we find some children learning only part of their language, or even for that matter, none at all? When we consider other types of learning, such as the learning of mathematics or history, we observe great ranges of differences in the amount learned, the rate of learning, etc. Yet we note that degrees of difference in intelligence, varying environmental conditions, parental input, levels of apparent "need," etc.—factors that indeed bear on the learning of other kinds of knowledge—do not seem to have as much effect, if any, on language acquisition. Moreover, we ask why all children pass through similar stages of development, all of which are systematic, though not necessarily at exactly the same rate. Thus, although we do find relatively minor variations in first language acquisition, they do not reflect differences in general intelligence, aptitude, training, need, or any other variable that would otherwise affect the rate and amount of what is learned.

In light of these observations, we come to the conclusion that children must be born with some highly specified knowledge concerning the structure of language in general to enable them to "discover" the particular grammar of their native language, based on what is less than ideal exposure to the primary data (samples of the language heard). There must be principles that are universal to language (*language universals*) that count as innate mental structure and thereby guide children to discover the particular rules of their native language.

It cannot be argued that children can infer the distinctions and rules of the language they hear, as psycholinguist David McNeill pointed out some years ago:

Nor ... could the child infer adult classes from parental speech without knowing in advance the range of possible distinctions. Parental speech offers useful guidance at this point *only* if this condition is met. An ability to infer something about language is the capacity to generalize a distinction once its relevance is noticed. We cannot conceive of it being a capacity to invent the distinctions themselves. A vast number of distinctions is possible in parental speech—only a few of which are important in English—and if a child had to invent rather than notice them, his chances of progressing to English would be microscopically small.[220]

In other words, as another leading linguist and philosopher, Jerry Fodor, has put it, "Any organism that extrapolates from its experience does so on the basis of principles that are not themselves supplied by its experience."[221] That is, the principles by which we interpret experience are supplied by the mind. In the case of language, innate principles of the mind—language universals—enable us to find the rules that underlie the samples of speech that we are exposed to as children.

Chomsky has strongly argued for this return to a rationalist philosophy of mind in his writings. Borrowing from the Cartesian notion of "innate ideas," Chomsky has shown how the seventeenth-century rationalists, inquiring into the nature of language as a "mirror of the mind," were led to the conclusion that language learning is a process determined by the *initial state of the mind*—biologically endowed with structures determining the general form of language ("innate ideas")—together with maturational development and interaction with the environment. Chomsky explains:

> Man has a species-specific capacity, a unique type of intellectual organization which cannot be attributed to peripheral organs or related to general intelligence and which manifests itself in what we may refer to as the "creative aspect" of ordinary language use—its property being both unbounded in scope and stimulus-free. Thus Descartes maintains that language is available for the free expression of thought or for appropriate response in any new context and is undetermined by any fixed association of utterances to external stimuli or physiological states (identifiable in any noncircular fashion).[222]

Seventeenth-century rationalism notes that knowledge arises on the basis of very scattered and inadequate data and that there are uniformities in what is learned that are in no way uniquely determined by the data itself. Consequently, these properties are attributed to the mind, as preconditions for experience. This is essentially the line of reasoning that would be taken, today, by a scientist interested in the structure of some device for which he has only input-output data. In contrast, empiricist speculation, particularly in its modern versions, has characteristically adopted certain a priori assumptions regarding the nature of learning (that it must be based on association or reinforcement, or on inductive procedures of an elementary sort . . .) and has not considered the necessity for checking these assumptions against the observed uniformities of "output"—against what is known or believed after "learning" has taken place.[223]

Empirical support for the *Innateness Hypothesis* is strong when we consider the facts of language learning presented on pages 92–96—that is, the universal acquisition of a highly intricate system of knowledge despite individual differences. Humans are "pre-wired" genetically to acquire a complex system of knowledge of a language if exposed to its use in a given speech community. As noted in chapter 5, only in cases of brain impairment or deformity will there be any real exception. If this innate mental structure responsible for language acquisition (*the language faculty,* or "language acquisition system")[224] is specified in terms of general properties of language, then the facts of child language acquisition previously given can be expected. The language faculty sets the boundaries on the kinds of grammars that are learnable, i.e, on the class of *possible* grammars, in a manner we will discuss in the next section.

Thus, to sum up, a rationalist approach views the complexity of the system acquired, the uniformity as well as the species specificity (more in a later chapter) irrespective of level of intelligence (barring a pathological condition), etc., as clear evidence that a specific kind of *inborn mental structure,* a *language faculty,* is the basis for language learning. Certainly, no empiricist theory can adequately account for the universal acquisition of such a highly intricate system of linguistic rules constituting competence.

To represent schematically the task of the child in

acquiring a first language, Chomsky has used the following "instantaneous"* language acquisition model, in which a Language Acquisition Device (LAD) accomplishes what the child does:

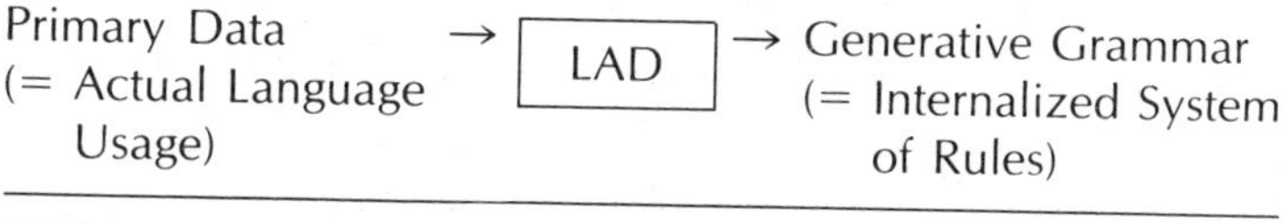

Theories Concerning Language Acquisition

Primary Data (= Actual Language Usage) → | LAD | → Generative Grammar (= Internalized System of Rules)

FIGURE 8. Language Acquisition Model

This diagram shows that with such varied input (grammatical and ungrammatical utterances from many sources) that the child is exposed to, he or she discovers (without instruction) the correct system of rules underlying that usage. The model draws attention to the fact that without a highly specific internal structure of the LAD (innate principles of the language faculty) there would be no way to explain the uniformity of acquisition of such an intricate system of rules.

We turn, now, to a brief discussion of the principles of the mind that the child is genetically equipped with in order to accomplish the task. These are described by linguists as general principles of language, called *language universals,* which determine the overall character of any human language.

In much of the current linguistic research, there is a great concern to discover and represent those properties of language that are supposed universal. These principles specify the *stock elements* of sound and meaning that the world's languages draw from (*substantive universals*), such as the *phonetic features* for possible *sounds* in language, and *lexical categories*

Linguistic Universals

---

*"Instantaneous" means that it is an idealization of the total acquisition process, which, as we have seen in chapter 5, involves a number of discernible stages. This model is described in Chomsky, *Aspects of the Theory of Syntax,* pp. 30–37.

like noun and verb; other universals determine the organizational structure of human languages (*formal universals*).[225] Formal universals provide *general conditions* on possible linguistic rules, including *phrase structure rules* and *transformational rules* (the two types of syntactic rules). Chomsky has emphasized the importance of this program for linguistics:

> The most intriguing of the studies of language structure are those that bear on linguistic universals, that is, principles that hold of language quite generally as a matter of biological (not logical) necessity. Given the richness and complexity of the system of grammar for a human language and the uniformity of its acquisition on the basis of limited and often degenerate evidence, there can be little doubt that highly restrictive universal principles must exist determining the general framework of each human language and perhaps much of its specific structure as well. To determine these principles is the deepest problem of contemporary linguistic study.[226]

While these abstract principles of language constitute what Chomsky has called a "mental organ" of language, their physical counterparts are not yet known: "This knowledge is in part shared among us and represented somehow in our minds, ultimately in our brains, in structures that we can hope to characterize abstractly, and in principle quite concretely, in terms of physical mechanisms."[227] Whatever directions future research may take, it seems clear that humans are endowed with a genetically determined program that delimits the nature and extent of what is acquired within the critical age before lateralization.

In his debate with famed psychologist Jean Piaget (see the following section, pages 102–107), Chomsky argued that in the absence of "relevant experience," there is reason to postulate innate knowledge to the learner, based on the abstractness of the system acquired.[228]

Consider his examples illustrating the *structure dependency* of language rules:

9) a) The man is here.
   b) Is the man here?
10) a) The man who is tall will leave.
    b) Is the man who tall will leave?
    c) Will the man who is tall leave?

Based on examples like those in (9), a child might construct a rule for the formation of questions, such as (11):

(11) Take the first occurrence of a (tensed) auxiliary like *is, will*, etc., and place at the beginning of the sentence.

However, this arithmetic type of procedure does not work in cases like (10); (10b) is not grammatical; rather (10c) is.

Thus the correct formulation of the question transformation must take account of the *phrase-structure* of the string; i.e., the noun phrase subject must be identified so that the correct auxiliary will be selected and moved around it. Thus, *the man who is tall* is the NP subject around which the auxiliary *will* is moved, thereby forming (10c). The question transformation, then, will be formulated approximately as follows:

(12) Move the first (tensed) auxiliary around the subject noun phrase (NP dominated by S).

We can see, therefore, that any correct formulation of the question (and any other transformation) must take account of the *phrase structure*, rather than the simple arithmetic arrangement of items in a sentence string.

This *structure dependency* condition is not something that is learned. No one ever has to tell a child to "look for" that type of rule, (12), as opposed to an arithmetic rule, (11), which is in fact a simpler type. How is it that we as humans select the more *complicated* type of linguistic rule? The answer is that such a condition on the kind of rule permissible in human language *must be* part of the innate knowledge that makes language acquisition possible in the first place. Thus Chomsky concludes:

> On the assumption of uniformity of language capacity across the species, if a general principle is confirmed empirically for a given language and if, furthermore, there is reason to believe that it is not learned (and surely not taught), then it is proper to postulate that the principle belongs to universal gammar, as part of the system of "pre-existent knowledge" that makes learning possible.[229]

Therefore, universal properties like the structure-dependency condition underscore the *abstractness* of human language. No one could conceivably learn, much less be made aware of, such principles that determine the nature of the linguistic rules comprising one's competence. Knowledge of language universals must accordingly be innate: the abstract mental faculty of language, which is "activated" in contact with normal language use, guides the child to "discover" the correct grammar of that language in an amazingly short period of time. Such a theory that the child constructs is presumed to have an organizational structure of the sort described in chapter 4, with phrase structure and transformational components, specifying deep and surface structures, respectively, etc.

To conclude this section, we again quote from Chomsky:

> A human language is a system of remarkable complexity. *To come to know a human language would be an extraordinary intellectual achievement for a creature not specifically designed to accomplish this task.* A normal child acquires this knowledge on relatively slight exposure and without specific training. He can then quite effortlessly make use of an intricate structure of specific rules and guiding principles to convey his thoughts and feelings to others, arousing in them novel ideas and subtle perceptions and judgments. For the conscious mind, not specifically designed for the purpose, it remains a distant goal to reconstruct and comprehend what the child has done intuitively and with minimal effort. Thus, language is a mirror of mind in a deep and significant sense. It is a product of human intelligence, created anew in each individual by operations that lie far beyond the reach of will or consciousness (italics added).[230]

Piaget's Constructivism

The influential Swiss psychologist Jean Piaget has written numerous works over several decades concerning the child's cognitive growth, including language development. While rejecting traditional empiricist views of learning, Piaget nevertheless is "anti-innatist" ("preformism" is Piaget's word for "innatism"). He states, "The functioning of intelligence alone is hereditary and creates structures only through an organization of successive actions performed on objects."²³¹ Thus Piaget sets forth a "constructivist" approach as an alternative to the "blank-slate" empiricist view on the one hand and the innatist view on the other.

Piaget's Theory of Language Development

Piaget links the development of language with that of general cognitive structures that have their origin in the sensorimotor period of the child's life, approximately the first year and a half.²³² What develops in the child are prelogical-mathematical structures from the child's interaction with objects in the environment. As Piagetian associate Sinclair-deZwart sums up the process:

> Piaget qualifies his epistemological theory as interactionist and biological. Knowledge is acquired through the subject's action upon, and interaction with, people and things. Action patterns become established, extended, combined with others, and differentiated under the influence of internal regulatory mechanisms; later, they become interiorized (i.e., mentally representable), and organized into group-like structures. Acting upon environment, rather than copying it or talking about it, is the source of knowledge. Language is only one way among others to represent knowledge. Representation in general does not appear until the end of the sensorimotor period (around the age of 1½) when direct-acting-on-objects has become organized in a first grouplike structure.²³³

Through *assimilation* the child "incorporates objects or events into actions and their existing knowledge. That is, new knowledge is always first interpreted by reference to existing knowledge."²³⁴ Fantasy and symbolic play exemplify this process. On the other hand, in the *accommodation* process, such as imitation, "internal structures are modified in accordance with environmental influences."²³⁵ These two basic processes are complementary in the child's *adaptation* to the environment, and through *auto-regulation* cognition develops—first prelogical, sensorimotor intelligence as a prelude to symbolic representation, then language and logical thought (after seven years). Each stage appears as an internalized structural reorganization of the previous one.[236]

Thus for Piaget there are no innate principles of the mind, except for these adaptive processes themselves, working within a biological organization.

Besides the Geneva group's working with Piaget, some psycholinguists have attempted to account for the emergence of syntax within the child, beginning around the age of two years, within Piaget's constructivist theory of cognitive development, including Roger Brown[237] and, more lately, David McNeill.[238]

Psycholinguists Donald M. Moorehead and Ann Moorehead summarize the relation between sensorimotor intelligence and the child's acquisition of sentencehood, in two stages: (1) early two-word utterances are said to derive from physical knowledge of external objects in the more advanced (sensorimotor) stages (e.g., the nominative existence and recurrence semantic relations) and (2) later telegraphic forms are said to result from the (pre-) logical-mathematical knowledge in the final stage, in which representation of previously acquired "internalized action schemas" take place.[239] Concepts such as agent-action, possession, location, and attribution thus are reflected in these more advanced telegraphic sentence types.[240]

Piaget himself has much to say about the socialization of the child and concomitant changes in the child's language development, starting from what he calls *ego-centric* to *socialized* speech around age seven.[241]

For Piaget, then, language in the child arises out of a series of increasingly complex responses to the environment, ultimately as internalized symbolic representation, once the prelogical mental structures are established by the end of the first year and a half of life.

Criticism of Piaget's Theory

Criticism of Piaget's constructivism has come from several sources, of which we will consider two. Psycholinguist-philosopher Jerry Fodor, for example, has argued that a child's acquisition of higher stages of development presupposes prior knowledge in the first place.[242] He begins by distinguishing *fixation of belief* (by experience) from *concept learning*, the latter being often confused for the former. He remarks that theories of fixation of belief must be "radically innatist" because there are no defensible theories of concept learning.[243] He points out that learning theories assume that "learning is a matter of inductive extrapolation, that is, of some form of nondemonstrative inference."[244] Thus one goes from some specific beliefs to general beliefs, involving the processes of hypothesis formation and confirmation.[245] However, the theory does *not* indicate where these beliefs come from in the first place. "In particular, it assumes as 'given' the 'criterial attributes' which form the hypotheses that are 'fixed' in the experimental situation."[246]

Fodor therefore argues that "a theory of how our beliefs are determined by our experiences is not a theory of the source of our inductive hypotheses. On the contrary, it presupposes the availability of such hypotheses. . . ."[247]

Fodor goes on to argue that Piaget's theory likewise presupposes a "given" at one stage in order for the next higher stage to be attained; i.e., to go from stage one, for example, to stage two by a learning process (hypothesis formation and confirmation) is impossible because a hypothesis at the higher level cannot be formed without "the conceptual apparatus available" at the lower stage. "It is *never* possible to learn a richer logic on the basis of a weaker logic."[248]

Thus in order for a child to attain to a higher (more complex) level of knowledge, as Piaget describes, the child must have the innate conceptual structures that, when *activated* by the environment and maturational development, will serve as the basis for further development. It turns out, then, that the acquisition of language, like other cognitive structures, crucially depends on the child's innate mental schema (prior knowledge) that he or she brings to the task.

The second criticism of Piaget's theory that we will consider comes from Chomsky.[249] Fundamentally, Chomsky argues against Piaget's claim that the sensorimotor intelligence provides the basis for the acquisition of language. Considering the abstractness of the system acquired, requiring the postulation of language universals, Chomsky points out that Piaget's claim borders on incredibility.[250] In other words, Chomsky's argument against empiricist views holds equally well for Piaget's constructivism, which, he argues, turns out to be a variant of empiricism.

Chomsky further mentions that in Piaget's theory, there is no principled explanation for the particular stages that a child indeed moves through. He comments:

> It is difficult to imagine what answer could be provided, apart from recourse to some assumption concerning maturation to a genetically determined target stage, at each point. And when such an assumption is made precise, it seems that it will express genetically determined aspects of human belief and knowledge that are far more intricate than the "elementary hereditary forms" that the Geneva school is willing to contemplate.[251]

In response, Piaget agrees with Fodor up to a point—that a particular cognitive structure contains something of the previous stage, but "*containing it not as a structure, but as a possibility*" (italics his).[252] Otherwise, Piaget maintains, by taking Fodor's (and Chomsky's) innateness theory, one would have to conclude that "nothing has ever been invented, that everything is always contained in the previous state."[253] To take the point to its extreme, Piaget argues that this position would allow simple organisms like viruses to have the innate potential to do higher mathematics.[254] Yet, Fodor disclaims such a conclusion: "The nativist isn't committed to saying that viruses know about set theory any more than he is committed to saying that viruses have legs; it hardly follows from the fact that viruses don't have legs that legs aren't innately specified."[255] Further, Fodor remarks that in concept learning (hypothesis formation and confirmation), the concepts of the previous state of knowledge have to be "*actually exploited to mediate the learning*," not simply "*potentially* accessible;"[256] otherwise, Fodor claims, the argument reduces to triteness—"whatever is learnable is learnable."[257]

We therefore accept these arguments by Fodor and Chomsky as valid criticisms of Piaget's thesis. These, in consequence, give credence to the innatist view of language acquisition.

Theories Concerning Language Acquisition

The Relation Between Language and Culture

Edward Sapir was a pioneer anthropological linguist and the first outstanding American descriptive linguist to move away from the prevailing behavioristic approach in the first half of this century. As a mentalist, Sapir stressed the psychological reality of linguistic units like the *phoneme*.[258] He viewed language primarily as "a vocal actualization of the tendency to see realities symbolically."[259]

As we discussed earlier (in chapter 3), Sapir, like other linguists of his time, emphasized that each language was a complete and complex symbolic system, regardless of whether the people who spoke it had a "primitive" culture or not. Yet, while recognizing this independence of language and culture, Sapir pointed out the interrelation between the two, based on the communicative use of the symbolic system as both a "socializing" and a "uniformizing" force.[260] That is, communicative language identifies groups and subgroups in society and provides psychological support for the various kinds of social solidarity and rapport among the members.

Nevertheless, Sapir also found in language use "the most potent single known factor for the growth of individuality," with particular qualities of expression constituting "complex indicators of the personality."[261] Today linguists are attempting to distinguish certain pragmatic aspects of language use that Sapir anticipated in his treatment.

Furthermore, Sapir pointed out that the notion of culture is not definable without the accompanying notion of personality: "The true locus of culture is in the interactions of specific individuals and, on the subjective side, in the world of meanings which each one of these individuals may unconsciously abstract for himself from his participation in these interactions."[262] So, he concluded, "The concept of culture, as it is handled by the cultural anthropologist, is necessarily something of a statistical fiction" apart from its true "metaphysical locus," which is in the human personality: "Every individual is, then, in a very real sense, a representative of at least one sub-culture which may be abstracted from the generalized culture of the group of which he is a member."[263]

Sapir, then, argued for an *integrative approach* to the study of culture and personality—from the "mother science" of social psychology ("which is not a whit more social than it is individual"),[264] as the following statement reveals:

> That culture is a super-organic impersonal whole is a useful enough methodological principle to begin with but becomes a serious deterrent in the long run to the more dynamic study of the genesis and development of cultural patterns because these cannot be realistically disconnected from those organizations of ideas and feelings which constitute the individual.[265]

In a similar fashion, Chomsky has argued for the inclusion of linguistics within cognitive psychology (and ultimately, human biology), since knowledge of a language is determined by innate structures in the mind.[266]

The issue of how the acquired knowledge involves a child's interaction with the environment revolves around the distinction between genetically determined cognitive development and "learning" from an essentially "blank slate," as we have already discussed. In his more recent publications, Chomsky has emphasized the interrelation between the language faculty and other cognitive structures in determining the overall development of linguistic knowledge.[267] Similarly, in light of Sapir's notion of the relatedness of the culture of a group and the personalities of the individuals in it, we might suppose that the range of acquirable cultural patterns is also genetically delimited, in some significant sense, with regard to what is knowable and transferable.

Thus, we can draw the conclusion that innate principles of the mind underlie not only the acquisition of language but also the development and transference of human cultural patterns.

While the organization of the mind ultimately determines the class of possible languages as well as cultures, it does not mean that these general constructs can be fully separated. Obviously, cultural knowledge is essentially transferred through language; that is, the semantic content of linguistic forms contains cultural information, such that within various communicative settings the propagation of this certain kind of knowledge is ensured.

In his early writings, Sapir noted that while language reflected culture, there was no one-to-one relation between the forms of a language and the patterns of the culture of its speakers: "It is difficult to see what particular causal relations may be expected to subsist between a selected inventory of experience (culture, a significant selection made by society) and the particular manner in which the society expresses all experience."[268]

Later, as an outgrowth of his emphasis on the psychological reality of linguistic structures, Sapir began to articulate a view that led to the formulation of *linguistic determinism*,* namely, that language structure shapes the thought patterns of its users:

> No matter how sophisticated our modes of interpretation become, we never really get beyond the projection and continuous transfer of relations suggested by our speech . . . . Language is at one and the same time helping and retarding us in our exploration of experience, and the details of these processes of help and hindrance are deposited in the subtler meanings of different cultures.[269]

This view was taken up by one of Sapir's students, Benjamin Lee Whorf, another anthropological linguist working among American Indian groups. The *Sapir-Whorf hypothesis* then took an even stronger form:

> Actually, thinking is most mysterious, and by far the greatest light upon it that we have is thrown by the study of language. This study shows that the forms of a person's thoughts are controlled by inexorable laws of pattern of which he is unconscious. These patterns are the unperceived intricate systematizations of his own language—shown readily enough by a candid

---

Linguistic determinism may be considered a strong form of *linguistic relativity,* the notion that each language is a separate system, without any necessary commonality with others, except what may be determined within language families.

comparison and contrast with other languages, especially those of a different linguistic family. His thinking itself is in a language—in English, in Sanskrit, in Chinese. And every language is a vast pattern-system, different from others, in which are culturally ordained the forms and categories by which the personality not only communicates, but also analyzes nature, notices or neglects types of relationship and phenomena, channels his reasoning, and builds the house of his consciousness.

This doctrine is new to Western science, but it stands on unimpeachable evidence. Moreover, it is known, or something like it is known, to the philosophies of India, and to modern Theosophy. This is masked by the fact that the philosophical Sanskrit terms do not supply the exact equivalent of my term "language" in the broad sense of the linguistic order. The linguistic order embraces all symbolism, all symbolic processes, all processes of reference and of logic.[270]

The Sapir-Whorf hypothesis was widely heralded in the 1940s and 50s. However, in the 1960s there were movements among structural and transformational linguists emphasizing the significance of universal properties of language. In addition to the transformational approach to the universality of abstract principles underlying human languages, there was a strong development among anthropological linguists, spearheaded by Joseph Greenberg, providing evidence for general tendencies among languages.[271] In summary, the strong Whorfian version of linguistic determinism has drawn criticism from various quarters, mainly for the lack of empirical support.[272]

One area of evidence that was offered in support of linguistic relativity and the stronger linguistic-determinism theories concerns the way a particular language divides up the spectrum in color terms and how this affects the speaker's perception of color. Advocates of linguistic relativism stressed that the division of color was arbitrary and culturally bound; advocates of the stronger linguistic determinism went even further to emphasize that such cutting up of the mass "out there" affected one's perception of what the colors were.

In recent years, evidence has been put forth by Berlin and Kay (1969) to show that there may well be underlying similarities across cultural boundaries concerning the "focal" meanings of color terms, regardless of the seeming differences in the main divisions of the basic colors.²⁷³ That is, while speakers of a particular language may have difficulty in determining the boundaries between two colors (indeterminate zones), they will agree on what is typical of each. Berlin and Kay demonstrate that such agreement exists across languages, leading them to conclude that there is a universal set of eleven basic color terms, with respect to which speakers of different languages make their particular distinctions. Moreover, they claim that the acquisition of color distinctions follows a well-defined order: black and white; red; green and yellow; blue; brown; purple, pink, orange, and gray.²⁷⁴

Based on this evidence for the universality of basic color terms, we therefore conclude that there is plausible support for the innateness hypothesis in semantic field theory that traditionally was held to be culture-bound.

Stronger theoretical argumentation that innate principles of the mine are responsible for the ways humans symbolize experience comes from ethnologist Dan Sperber.²⁷⁵ Sperber argues that since anything in the environment can be symbolized, there is no set of rules determining the class of symbolic concepts; thus no set of principles may be inculcated in ("taught" to) a child. Rather, there is an inborn capacity to symbolize, and this capacity must therefore be distinguished from linguistic knowledge, because the latter *is* determined by a set of rules. There being no "grammar of symbolism," the capacity to invent symbols that appear culturally bound must consequently be based in innate "aptitude."²⁷⁶

Conclusion

The facts of language acquisition have shown the rationalist view to be compelling: the complex nature of the object that is uniformly acquired (linguistic competence) leaves little doubt that one could ever accomplish such a task without being genetically equipped with specific (though abstract) principles determining its essential character. What would weaken this argument would be evidence that the apes can attain the same kind of knowledge (shown through symbolic behavior of different sorts) resulting from the intervening efforts of psychologists.²⁷⁷

chapter seven

Animal Systems of Communication and Human Language

This chapter generally compares animal systems of communication with human language. The distinguishing characteristics of each system are presented. The authors demonstrate that human linguistic ability is not merely due to higher intelligence but to a specific type of mental organization.

In chapter 3 we saw that human language is open-ended, or unbounded in scope: there are an unlimited number of signals (sentences) expressing an unlimited number of messages that native speakers are capable of producing and understanding. In the normal use of

A General Comparison of Human Language and Animal Systems of Communication

language, such novel sentences are appropriate to the occasion (thus forming connected discourse) and yet are stimulus-free, i.e., not restricted by external conditions or internal states of the user. It must also be remembered that human language is discrete—composed of perceivable units of sound (phonemes) and form (morphemes) comprising words that make up constructions, resulting in sentences. Nevertheless, while human language is discrete, it is unbounded in scope.

In contrast, Chomsky has pointed out, natural animal communication systems are either discrete and bounded, or nondiscrete and unbounded:

> Every animal communication system that is known . . . uses one of two basic principles: Either it consists of a fixed, finite number of signals, each associated with a specific range of behavior or emotional state . . . or it makes use of a fixed, finite number of linguistic dimensions, each of which is associated with a particular nonlinguistic dimension in such a way that selection of a point along the linguistic dimension determines and signals a certain point along the associated nonlinguistic dimension.[278]

Using Chomsky's framework, Akmajian, Demers, and Harnish place bird calls and songs in the discrete-bounded category, and the bee dance in the nondiscrete-unbounded category.[279] Bird calls, for example, are "sound patterns consisting of single notes. . . ."[280] Thus the calls are discrete—notes that signal a limited number of messages (i.e., they are bounded). Yet each discrete signal has graded possibilities;[281] e.g., the louder the call, the greater the indication of danger from a predator.

On the other hand, the fascinating communication system of honeybees (known as "dancing"), such as the tail-wagging dance, is nondiscrete and unbounded, as described here:

> The tail-wagging dance conveys the following information: the *orientation* of the straight-line portion of the dance communicates the *direction* that the bees must fly with respect to the position of the sun; the length of *time* spent during the tail-wagging portion of the dance communicates the *distance* that must be flown; and, finally, the general level of *excitation* during the dance communicates the *richness* of the source of food.[282]

However, as Akmajian and his colleagues carefully point out, the unboundedness of the bee dance is only *trivially similar* to that of human language; the unbounded number of messages are all related to the source and quality of food. Human language, by contrast, is unbounded in a significantly different way: there is virtually *no limit* to either the content or the form of message that may be signaled. The point of difference, therefore, is, as Chomsky emphasizes, "one of an entirely different principle of organization."[283] Similarly, linguist Ronald Langacker observes that the difference between the nature of the unboundedness of an animal system and that of human language "would appear to be much more impressive than the sole similarity . . . namely, that, like human language, some animal communication relies on fixed systems of signals."[284] Moreover, he points out that there is a "vastly greater structural complexity of the signals of a human language" as compared with the lack of internal structure of the signals in animal systems.[285] In short, Chomsky concludes, "When we ask what human language is, we find no striking similarity to animal communication systems."[286] (Further discussion along these lines will be presented in chapters 8 and 9.)

Animal Systems of Communication and Human Language

Natural Communication Systems of the Great Apes

What, then, of the natural communication systems of the primates, including chimpanzees and other great apes? First it should be noted that they make use of both visual and auditory signals within remarkably structured social orders; thus postures and gestures as well as calls are included.[287] Akmajian,[288] and other linguists such as Langacker, characterize all the primate modes (calls included) as discrete and bounded, along with bird songs and calls; again, each call having graded possibilities.[289]

Since the rhesus monkey has been one of the most carefully studied primates (though it isn't an ape), we reproduce on page 116 a tabular description of nine harsh, antagonistic calls of the rhesus.[290]

The higher pitched acoustic signals described in Chart 3 are due to a primate's somewhat different

Call	Description	Situational Context
Roar	Long, fairly loud noise	Made by a very confident animal, when threatening another of inferior rank
Pant-threat	Like a roar, but divided into "syllables"	Made by a less confident animal, who wants support in making an attack
Bark	Like the single bark of a dog	Made by a threatening animal who is not aggressive enough to move forward
Growl	Like a bark, but quieter, shriller, and broken into short units	Made by a mildly alarmed animal
Shrill-bark	Not described	Alarm call
Screech	Involves an abrupt pitch change, up then down	Made when threatening a higher-ranked animal
Geckering screech	Like a screech, but broken into syllables	Made when threatened by another animal
Scream	Shorter than a screech and without a rise and fall	Made when losing a fight and being bitten
Squeak	Short, very high noise	Made by a defeated and exhausted animal at the end of a fight

CHART 3. Rhesus Monkey Calls

Source: Table 7, "Rhesus Calls" (after Marler and Hamilton, 1966, and Rowell and Hinde, 1962) from *The Acquisition of Language: The Study of Developmental Psycholinguistics* by David McNeill.
Copyright © 1970 by David McNeill. Reprinted by permission of Harper & Row, Publishers, Inc.

configuration of the upper vocal tract (especially the length of the pharynx), which serves as a "filter" of the sound waves produced by the vocal cords.[291] (See page 117.)

Akmajian et al. cite primatologist Thelma Rowell with respect to the fact that "communication by monkeys is not qualitatively different from that of other animals."[292] Primatologist Peter Marler identifies thirteen different sounds in chimpanzees, with gradations within each unit.[293]

Thus as we approach the evaluation section (in the

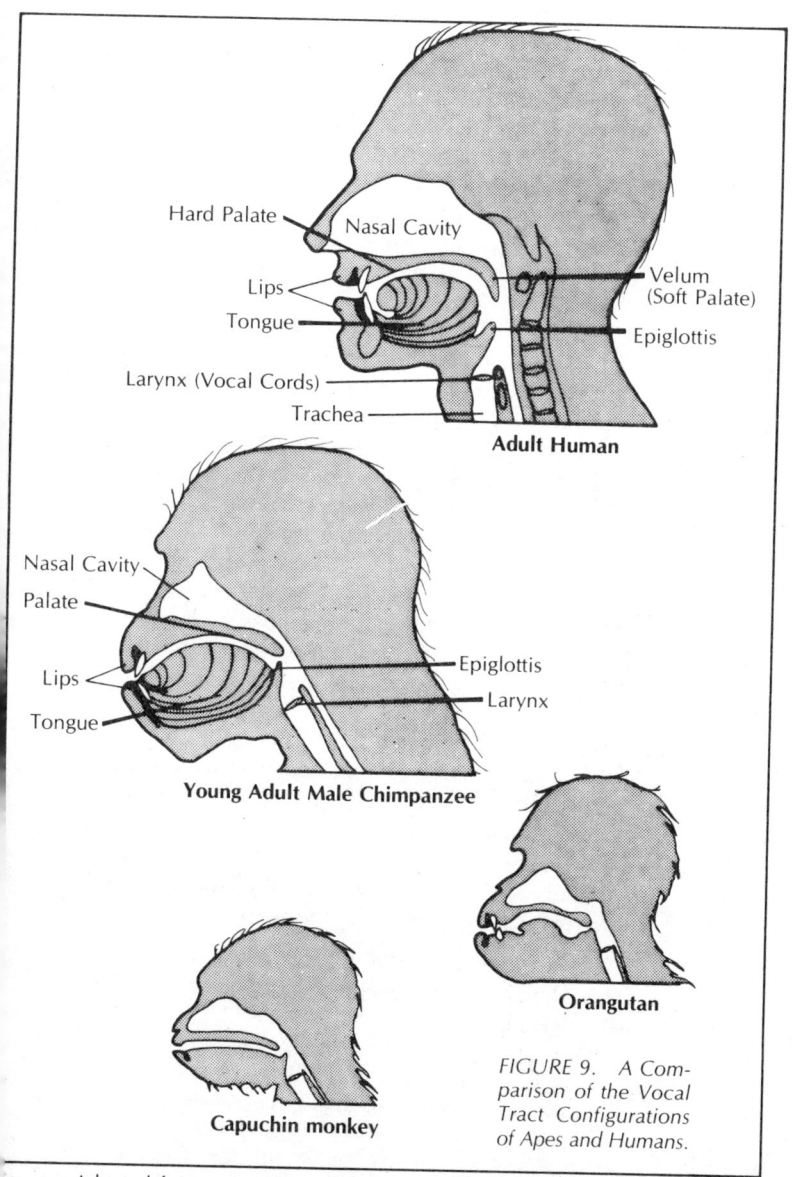

FIGURE 9. A Comparison of the Vocal Tract Configurations of Apes and Humans.

Source: Adapted from V. E. Negus, *The Comparative Anatomy and Physiology of the Larynyx* (Darien, Conn.: Hafner, 1979). By permission of William Heinemann, Ltd., London.

next chapter) of the attempts to teach chimpanzees a languagelike system, we note the conclusion made by Akmajian and his colleagues concerning the natural communication systems of the great apes: "The gap between primate communication systems and human linguistic communication is huge."[294] This observation is echoed by other linguists, following the definitive remarks by Chomsky concerning the overall relation of human language to animal systems in light of evolutionary theories[295] (see chapter 9).*

General Intelligence Versus Specific Mental Faculty of Language

In chapter 6 we observed that the spontaneous and universal acquisition of such a complex and abstract system of linguistic competence was compelling evidence for the innateness hypothesis. Humans are genetically endowed with a unique "mental organ"[299] that gives rise to the development of an internalized system of competence upon exposure to apparently fragmentary primary data (actual usage of adults and other children). Thus, as Chomsky has strongly argued, "As far as we know, possession of human language is associated with a specific type of mental organization, not simply a higher degree of intelligence."[300]

Let us recall the argument in chapter 6 (Linguistic Universals, pages 99–102) concerning the lack of correlation between language acquisition and general levels of intelligence. If the acquisition of language were a product of general intelligence (and thus essentially learned as any other kind of knowledge), then we would expect the following:

1. different levels of acquisition, dependent on the level of intelligence, other supposed factors being equal

---

*A popular analysis of various (fourteen) defining characteristics of human language was devised by linguist Charles Hockett[296] and has been used to compare language with animal communication systems, e.g., by ethologist William H. Thorpe.[297] Akmajian et al. appraise its usefulness for comparison purposes in view of its emphasis on the acoustic characteristics of human speech.[298]

2. language form to be in a direct, one-to-one relation with meaning; i.e., one sentence form would have only one meaning, and vice versa
3. the possible sentences in the language to be finite in number, since the child "would have no reason not to believe that a language was just a finite list of utterances that had to be memorized"[301]

Instead, we find just the opposite to be true in each case: all normal children acquire a full system of competence; language is a highly complex system, as evidenced in the relations of ambiguity, paraphrase, and opacity (cf. chapter 3); and the number of sentences in a language is infinite.

In the following chapter, we will consider this issue further with respect to the higher apes, which have clearly demonstrated surprising levels of intelligence, as the man-ape communication experiments have shown (cf. chapter 2).

A Review of Language Characteristics as Preparation for Evaluating Primate Attempts at Language Learning

As a means of refocusing on the essential features of human language that have been presented from chapter 3 on, we list below twelve characteristics that will also serve as a basis for evaluating the efforts of the higher primates to learn languagelike systems, as described in chapter 2.

With respect to the nature, use, and propagation of human language, then, we may list the following characteristics:

Its Nature

1. Discrete as a sound-signaling system, speech being the primary manifestation of language, except when abnormal conditions require the use of a gestural sign language.
2. Symbolic (and arbitrary) with respect to the meaning of the signals; language thus manifests a duality of sound and meaning.
3. Rule-governed, or systematic, with respect to the sound-meaning correspondence at all levels.

4. Compositional with respect to the determination of meaning from the constituents in their syntactic constructions.
5. Complex with respect to the sound-meaning correspondence; relations of ambiguity, synonymy, and opacity demonstrate this characteristic.
6. Displacing; that is, capable of expressing nonvisible, nonsensory, imagined, and otherwise abstract information.

Its Use

7. Unbounded in scope (open-ended), capable of expressing novel thoughts in novel ways (new sentence forms are limitless).
8. Independent of stimulus control, whether external conditions or internal states (see #6 above).
9. Suitable for contextualized communication; a native speaker may express novel sentences appropriate to the topic of discussion.

Its Propagation

10. Independent of need satisfaction; a child does not acquire a language because of any fulfillment of particular physical or emotional needs; variations in need level have not been shown to have any significant effect on the acquisition process.
11. Spontaneously acquired; for all normal children, mere exposure to its use is sufficient for language acquisition.
12. Culturally transmitted; while being genetically determined, language is learnable from the speech community, re-created by each individual in the social setting.

This summary prepares us to examine the communication experiments with the primates because most of the psychologists involved have, at one point or another, made claims concerning the similarity of their subjects' behavior to human language usage (at different stages).

chapter eight

Evaluation of Ape Communication With Man

The strong claims made by the researchers of ape-human communication entail that the primates must be able to create sentences in much the same way as humans do. Most linguists are skeptical of that claim and here the authors show why. The claims are then reevaluated and compared to human linguistic ability.

In light of what has been presented concerning the nature, use, and acquisition of human language in chapters 3–7, we now turn to the major question concerning ape communication behavior as elicited by the researchers: Can apes learn and use a communication system resembling human language? Or to put it otherwise, *Can apes create sentences?*

Review of Strong Claims Concerning Ape Symbolic and Grammatical Usage

To answer this question, we will evaluate, on the basis of the twelve criteria of human language presented in the preceding chapter, the claims made by the researchers of ape-human communication. We will also present arguments by many of the primary researchers themselves concerning their own previous work and that of others. This body of reevaluation literature has become sizeable in the last several years and has focused on both the claims themselves and the methodological assumptions and specific techniques used in the experiments.

Below is a review of the strong claims made by researchers who, at one point or another, have argued that there is no great qualitative difference between ape communication behaviors and human language:

1. R. A. and B. T. Gardner:

> In the further development of her combinations, in her replies to *wh*- questions, in her use of negatives, prepositions, and locatives, Washoe compared favorably with children at [Roger] Brown's Stage III and beyond.[302]

> We can marshal a very large body of empirical evidence and independent linguistic analysis, all agreeing with the conclusion that the communicative gestures of our chimpanzees would be called signs if they were used by human children and that the chimpanzees use signs in a rudimentary, childish form of Ameslan.[303]

> Behavior that is at least continuous with human language can be found in other species.[304]

2. Anne and David Premack:

> It seems clear that language is a general system of which human language is a particular, albeit remarkably refined, form.[305]

> Compared with a two-year-old child . . . Sarah holds her own in language ability. In fact, language demands were made of Sarah that would never be made of a child.[306]

> The test with the compound sentence [cf. chapter 2, p. 28] is of considerable importance because it provides the answer to whether or not Sarah could understand the notion of constituent structure: the hierarchical organization of a sentence. . . . If Sarah were capable only of linking words in a single chain, she

would never be able to interpret the compound sentence with its deletions. The fact is that she interprets them correctly.[307]

3. Duane Rumbaugh (with T. V. Gill):

An important development in the experiments reported here was Lana's acquisition of many critical linguistic-type skills for which she had received no specific training. Unquestionably she would never have acquired these skills spontaneously had she not received prior, very specific training in certain language fundamentals; nevertheless, having received such training, she showed a readiness to expand her ability to use linguistic-type communication in a number of significant, and, to her, novel directions.[308]

We believe that the success Lana has had so far in acquiring linguistic-type skills supports our view of language—that the foundations of language are to be found in the processes of intelligence. Man's outstanding ability to use language is, in part, a function of his high intelligence. We suggest that chimpanzees do not use public language in the field because they lack sufficient intelligence to agree upon the meanings of things that might serve as words and otherwise to develop language-type communication systems.[309]

4. Francine Patterson:

Both species of ape [chimpanzee and gorilla] have exhibited close parallels to human children with respect to the development of semantic relations in early language.[310]

In Koko's overgeneralizations, there is evidence that as she acquired words, she was actively involved in investigating the rules that governed her new tool.[311]

The Gardners, Roger Fouts, and I still believe there is sound evidence to support the claim that apes use sign language spontaneously, appropriately, and creatively. . . .[312]

. . . language is no longer the exclusive domain of man.[313]

Evaluation of Ape Communication With Man

Linguists' Skepticism

Concerning the validity of these and other claims, it must be said at the outset that linguists in general have remained skeptical. This dim view concerning the primates' supposed ability to learn a form of human language comes out of a context in which animal systems have been found to be different in *kind*, not just in degree. It must also be remembered that linguists' discoveries concerning the systematic and complex nature of language have prompted a postulation of a richly endowed mental structure that makes language acquisition possible. Thus Chomsky's position has been unwaveringly supportive of the Cartesian view, as he states in *Language and Mind:*

> Anyone concerned with the study of human nature and human capacities must somehow come to grips with the fact that all normal humans acquire language, whereas acquisition of even its barest rudiments is quite beyond the capacities of an otherwise intelligent ape—a fact that was emphasized, quite correctly, in Cartesian philosophy.[314]

With respect to modern attempts to refute this view of the *qualitative* difference between human language and animal systems, Chomsky further remarks:

> It is widely thought that the extensive modern studies on animal communication challenge this classical view; and it is almost universally taken for granted that there exists a problem of explaining the "evolution" of human language from systems of animal communication. However, a careful look at recent studies of animal communication seems to me to provide little support for these assumptions. Rather, these studies simply bring out even more clearly the extent to which human language appears to be a unique phenomenon, without significant analogue in the animal world.[315]

Since Chomsky wrote this (in the late 1960s) before the vast amount of research beginning with Washoe was done, it would be interesting to see whether he has altered his view concerning the recent attempts at teaching apes. In a 1976 lecture at the Washington School of Psychiatry, Chomsky reiterated his earlier position:

> Can we expect to find, in other organisms, faculties closely analogous to the human language capacity? It is conceivable, but not very likely. . . . It is difficult to imagine that some other species, say the chimpanzee, has the capacity for language but has never thought to put it to use. . . . The interesting investigations of the capacity of the higher apes to acquire symbolic systems seem to me to support the traditional belief that *even the*

most rudimentary properties of language lie well beyond the capacities of an otherwise intelligent ape (italics added).³¹⁶

Finally, in a 1978 paper delivered at the annual meeting of the American Association for the Advancement of Science, Chomsky reaffirms his earlier position in answering the Gardners' claims:

> Summarizing, recent work seems to confirm, quite generally, the not very surprising traditional assumption that human language, which develops even at very low levels of human intelligence and despite severe physical and social handicaps, is outside of the capacities of other species, in its most rudimentary properties, a point that has been emphasized in recent years by Eric Lenneberg, John Limber, and others. The differences appear to be qualitative; not a matter of "more or less," but a different type of intellectual organization, so it appears.³¹⁷

While not discounting their usefulness, Chomsky has argued that the attempts to induce apes to learn a symbolic system could confuse an observer to think that they are using a system like human language, in much the same way as a person who jumps is really thought to be flying, although to a lesser extent.³¹⁸

We come now to the question why Chomsky and other linguists have taken a dim view of the claims made by the above comparative psychologists. We approach the answer by recalling the distinguishing features of human language summarized in chapter 7. Can ape signing behavior and other such accomplishments be characterized by these criteria, even in embryonic form (since no one claims that an ape has fully exhibited a system of adult linguistic competence)?

After that we will review the data summarized in chapter 2 in light of these criteria of language, but not as a debate between linguists on one side and comparative psychologists on the other.* In recent years during the man-ape language debate, there has been an interesting body of literature from a number of the

---

*We are not attempting to put all linguists in the same category as transformational grammarians like Chomsky, since some have a more wait-and-see attitude.

comparative psychologists themselves and from various linguists that has cast serious doubt on the validity of the strong claims mentioned at the beginning of this chapter.

These reevaluations have dealt with the teaching and testing procedures used, as well as with the recording and interpretation of the data. Most of the researchers mentioned in chapter 2 have been critical of both the work of others and their own; some have even reversed their original conclusions, as we will see. Thus the debate has considerably widened to include the views of certain comparative psychologists who have taken another look at their own work.

Background: "Rich Interpretation" in Child Language Research

The Interpretation Problem: "Rich Interpretation" and Overattribution

While the ape-man communication research was picking up its pace around 1970, the psycholinguistic literature concerning child language acquisition had already made a strong impact in various circles. Especially significant were the findings concerning the character of the emerging grammars of young children, using transformational-generative linguistic models, as discussed in chapter 5. In view of the child's soon-to-be-acquired systems of adult competence, psycholinguists like Bloom (1970), Schleshinger (1971), and Brown (1970, 1973) employed a method of "rich interpretation" of early child utterances, based on observed parental expansions of children's telegraphic speech. By "rich interpretation" we mean the attribution of contextual information (including semantic relations inferred from adult equivalent sentences) to the child's utterances lacking the syntactic devices needed to formalize semantic notions (such as agency and possession).[319]

Justification for such an analysis can be found in the following facts: (1) a child's telegraphic utterances exhibit similarities in basic word order with corresponding adult sentences (see the section on developmental stages in a child's language acquisition, pages 76–88); (2) a child can understand adult sentences exhibiting such semantic relations; and (3) a child's own utterances become progressively more complex;[320] and, furthermore, (4) a child's errors in overgeneralizing are especially revealing of his or her propensity to look for and apply linguistic rules. Therefore, the attribution of contextually related semantic information to a child's telegraphic forms—no doubt related to other cognitive development—seems justifiable on the above grounds.

"Rich Interpretation" in Ape Communication Research: Overattribution Versus Parsimonious Interpretation of Sequences

From their early work with Washoe, the Gardners, likewise applied the method of rich interpretation to the sign combinations they recorded. They assigned the same semantic relations to Washoe's sign combinations as Roger Brown did to the early utterances of children (see table 3, page 36). They refer to responses in *obligatory contexts*—"experimenter-determined situations, in which context plus a particular question determine the correct response."[321] (See table 5, page 128.)

Herbert Terrace appears to have been among the first of the comparative psychologists to express serious doubt about the early claims by the Gardners, David Premack, and Duane Rumbaugh concerning their subjects' reputed linguistic abilities. He criticized his fellow researchers for what he and his Project Nim collaborators (including psycholinguist Thomas Bever) would later deem *overattribution*, involving an unjustifiable use of the method of rich interpretation. Moreover, after reexamining his own Project Nim, Terrace found further evidence for the inappropriateness of a rich interpretation of ape signing and other elicited behavior.[322]

Terrace and his colleagues point out three sources of this phenomenon of "unvalidated interpretation" that can creep into an analysis of such data: (1) the subjective nature of semantic interpretation (for example: the assignment to a noun of a semantic role like *agent* or *benefactor*), (2) the relatively small number of lexical items (expressing the various semantic functions) that enter into the combinations under study, and (3) an underestimation of the extent of the teachers' modeling in shaping the ape's sequences.[323]

Hence, these researchers argue that a more parsimonious interpretation should be sought first: "A rich interpretation of a sequence of signs as a sentence

TABLE 5
Parallel Descriptive Schemes for the Early Combinations of Children and Washoe

Brown's scheme for children[a]		The scheme for Washoe	
Types	Examples	Types	Examples
Attributive: (Ad + N)	Big train; Red book	Entity and attribute:	Drink red; Comb black
		Animate and trait:	Washoe sorry; Naomi good
Possessive (N + N[b])	Adam checker; Mommy lunch	Entity and possessor:	Clothes Mrs. G.; You hat
		Entity and possessive:	Baby mine; Clothes yours
Locative: (N + V)	Walk street; Go store	Action and locative:	Go in; Look out
Locative: (N + N)	Sweater chair; Book table	Action and location:	Go flower; Pants tickle[c]
		Entity and locative:	Baby down; in hat[d]
Agent and action: (N + V)	Adam put; Eve read	Agent and action:	Roger tickle; You drink
Action and object: (V + N)	Put book; Hit ball	Action and object:	Tickle Washoe; Open blanket
Agent and object: (N + N[b])	Mommy sock; Mommy lunch	(Not applicable)[b]	
(Not applicable, see text)		Appeal and action:	Please tickle; Hug hurry
		Appeal and entity:	Gimme flower; More fruit

[a]Brown, "The First Sentences of Child and Chimpanzee," pp. 85–101.
[b]Indicates types classified in more than one way in Brown's scheme and only one way in our [the Gardners'] scheme.
[c]Answer to question, "Where tickle?"
[d]Answer to question, "Where brush?"
Source: Gardner and Gardner, "Comparative Psychology and Language Acquisition," in Sebeok and Umiker-Sebeok, *Speaking of Apes*, p. 289.

[with various semantic functions attributed] has to be supported by clear demonstrations that such sequences cannot be produced by nongrammatical processes."[324] In other words, as linguists Sebeok and Umiker-Sebeok have argued, "Ocham's razor"—a principle of economy of explanation (parsimony) employing the fewest assumptions—should be used in interpreting the significance of the ape research.[325]

Accordingly, a rich interpretation of a child's telegraphic utterances is justifiable—though psycholinguists have cautioned against its overuse—because of the eventual acquisition of a full system of competence and the otherwise great difficulty in accounting for the child's emerging capacity to produce and understand sentences. As Terrace and his associates state, "Explanations of their [children's] utterances that are not based upon some kind of grammar become too unwieldy to defend."[326] However, in the case of the apes, a rich interpretation is not easily justified on this basis because it cannot be *assumed* that an ape will eventually develop a system of linguistic rules equivalent to grammatical competence in humans.

Not surprisingly, the Gardners have countered that Terrace and colleagues have applied "a rubber rule," claiming that if the method of rich interpretation is applicable in the case of child language learning, why not with ape language learning?* Seidenberg and Petitto answer this charge by stating that the Gardners and Patterson have not presented enough data "on which to base any interpretation."[327] Believing this to be the essential problem in the differences in interpretation, they also stress that "Patterson and the Gardners have consistently misinterpreted the 'method of rich interpretation' to mean, 'apply the richest interpretation to the data' without evaluating weaker alterna-

---

*It seems ironic that behaviorists like the Gardners would take a mentalist approach to data interpretation (i.e., attributing mental constructs of semantic categories for apes to attempt a comparison with children's language development), only to cry unfair at descriptions of apes' behaviors without such otherwise unwanted constructs.

tives."[328] The Rumbaughs now agree, stating, "The researchers who began to teach apes language did so with an inherent bias toward a set of presumptions formerly reserved for children."[329]

Consequently, some of the primary researchers themselves have taken the position that they must first look for nonlinguistic processes that do not rest on the strong assumption that an ape is using linguistic rules in creating sentences.[330] For example, Terrace and associates have reconsidered Nim's regularities in two-sign sequences involving a small number of signs and semantic rules. They conclude that, rather than attributing the knowledge of semantic rules to Nim, a simpler explanation would be that Nim made use of idiosyncratic "lexical position habits," i.e., displaying particular order preferences of various signs.[331]

Another more parsimonious (simpler) interpretation of many of Nim's other combinations is that Nim imitated his teachers a lot.[332] Having reached this conclusion after a careful analysis of the videotapes, Terrace and colleagues explain, "An analysis of video tapescripts revealed yet another spurious source of the semantic look of Nim's combinations: the extent to which Nim's utterances were initiated by his teacher's signing and were imitations of his teacher's preceding utterance."[333]

This admission concerning the role of imitation will be taken up in the next section because it concerns another major area of difficulty in interpretation in the ape communication research. These same researchers[334] and other primary researchers like the Rumbaughs have more recently emphasized that there is lack of systematic data on which to make any strong claims.[335] The point here is that a more parsimonious explanation for a large number of the signing apes' sequences is that they are due to *imitation* (and *reduction*) of their teachers' adjacent utterances and promptings.

Overattribution Concerning the Linguistic Status of the Forms

A related matter concerns the interpretation of the form of the response behavior itself. Seidenberg and Petitto have argued that "it is inaccurate to term their behavior 'signing in ASL.' Of course, the apes might have acquired facility in a sign language that was not ASL, e.g., a pidgin sign."[336] They base their conclusion on three facts. First, the apes were exposed to modified sign forms, sometimes due to some lack of manual dexterity,[337] as well as to the rather limited universe of discourse (mostly concerning the gratification of physical needs). Second, many of the primary teachers were not fluent in ASL to begin with. Third, the apes have not clearly demonstrated the grammatical structure of ASL, more of which will be discussed below.

Seidenberg and Petitto thus conclude that a more parsimonious or "weaker claim would be that the apes learned lexical items of ASL—i.e., the citation forms of signs—but none of its grammar."[338]

Commenting on Washoe's use of ASL, Chomsky concludes similarly: "To use symbols correlated by the investigator with symbols of Ameslan (even if the correlation is one of visual matching) is not necessarily to use Ameslan, as they [the Gardners] mistakenly conclude."[339] Then Chomsky points out a more general tendency that gives rise to overattribution—*mistaking certain similarities for essential sameness:* "To act in a manner resembling the early stages of the manifestation of some system of another species is not necessarily to be producing incipient manifestations of this system, obviously. Once again, we are back to the question whether certain analogies are useful and suggestive."[340]

The signing apes are not the only subjects scrutinized in this regard. The Gardners criticize David Premack and Duane Rumbaugh (in their earlier work) for attributing their respective ape subjects with understanding certain meanings (which these researchers invested) in successfully completing various responses to gain a reward.[341] Highly conditioned forced-choice tests (as with Sarah particularly) could be easily solved by the rote-memory capacity of a chimpanzee. Similarly, numerously repeated trials of drill or use of a small number of keys and stimulus arrays could be well within the constraints on the rote memory of a chimpanzee.[342] (See also Problem Solving, Not Language Learning, pp. 144ff.) The Gard-

131

Evaluation of Ape Communication with Man

ners correctly judge Premack's and Rumbaugh's lack of evidence for true symbolic understanding with their ape subjects: "With caged subjects and forced-choice tests, the results that Premack and Rumbaugh, et al., have presented thus far are more parsimoniously interpreted in terms of such classic factors as Clever Hans cues, rote memory, and learning sets."[343]

Ironically, though, the Gardners' own work would not be exempted from the criteria with which they evaluated the others, as Chomsky later remarked,[344] despite their claim to the contrary: "With our free-living subjects and naturalistic conditions, we developed testing procedures that rule out these alternative interpretations."[345]

Terrace and his colleagues go so far as to say that since "Sarah's and Lana's multisign utterances are interpretable as rotely learned sequences of symbols arranged in particular orders . . . it seems more prudent to regard the sequences of symbols glossed as *please, machine, Mary, Sarah,* and *give* as sequences of nonsense symbols."[346]

Rumbaugh and his associates[347] later accepted these conclusions of the Gardners and the Terrace group concerning the lack of evidence for true symbolism in a reward-reinforcement test.[348] They also argued that symbols should not be considered real words unless they are "used properly in a setting which requires that animal to make an accurate and specific evaluation of his environment and then use these symbols accurately and as communicative referents."[349] Thus they conclude that "the crucial issue is knowledge of the various contexts and forms of usage, not number of days on which the word was produced [referring to the Gardners' criterion], nor an 80% score on a particular two choice word naming task [referring to Premack's criterion]."[350]

Instead of being understood as symbolic referents for things, the Rumbaughs say that such things may be "highly generalized motor patterns which are learned as appropriate and desirable behaviors to emit in a variety of contexts and are therefore not necessarily communicative."[351]

In a later study (1980) Rumbaugh and his colleagues concede that "no theoretical framework has been developed within the field of animal psychology to distinguish between conditioned discriminative responses and symbolic representational responses."[352] They later make this comparison of apes and children:

> Apes then, like children, learn to use symbols as part of social-interaction routines. They are able to discern various sets of circumstances in which the production of particular symbols is deemed appropriate and results in obtaining a goal. They, like children, also learn to initiate these social interaction routines by producing symbols. Unlike children, however, apes do not seem to have moved beyond this point. *To date, there is no evidence that Washoe, Sarah, Lana, Koko, or Nim achieved symbolization proper* (italics added).[353]*

Overattribution Concerning "Aptness and Creativity" of Expression

In reference to Patterson's and the Gardners' claims, Pettito and Seidenberg ask how an ape can be said to know the meaning of such abstract words as *sorry, please, happy, sad, good, bad, silly, funny,* etc.[354] "Each of these attributions entails strong claims about the apes' cognitive capacities—their perception of the world, ability to make comparative judgments, awareness of self and others, conceptual skills, ability to consciously monitor their own behaviors, and the like—which are vastly underdetermined with respect to the evidence provided."[355] Therefore, it appears to be the investigators' *intentions* that unwittingly creep into the interpretations of the signs or symbol manipulation.[356]

In a similar vein, the Sebeoks seriously question Patterson's and the Gardners' accounts of the seeming aptness and innovativeness of the responses of their

---

*Nevertheless, the Rumbaughs claim that two other subjects, Sherman and Austin, have achieved a level of symbolic representation missing in other studies. Following certain training, these chimps were able to categorize various arbitrary lexigram symbols as either "food" or "tool." The Rumbaughs view this achievement—though quite limited in scope—as a prerequisite for any real language learning. ("Do Apes Use Language?" p. 60 and references cited there.)

subjects, when, in simpler terms, the animals have responded inappropriately. The Sebeoks remark that "human trainers appear all too willing to stretch their imaginations in order to make the animals' performance 'fit' conversationally. . . . Thus, anomalous chimpanzee or gorilla signs may be read as jokes, insults, metaphors or the like. . . ."[357] The point is not that man alone has a corner on deception but that the tendency for a researcher to read a linguistic ability into an otherwise unaccountable behavior is unwarranted. Consider further examples, given by Patterson. (See table 6, page 135.)

In looking over some of Patterson's interpretations, one wonders whether Koko knew what she was signing. What sense of *darn*, for example, could she have had? How could Patterson have conveyed such meaning? What sense of *devil* could she have understood—what kind of "theology" was inculcated into her? Notice how Patterson adds (derivational) morphological information to Koko's supposed sign for *stink*. Numerous questions along these lines may be raised in each of the instances cited.

In one of the most-quoted examples of an ape's creativity, Roger Fouts richly interprets Washoe's identification of a swan as a "water bird," in response to his question "What's that?" Terrace and various other researchers[358] have stated that a simpler explanation would be that she was identifying first the water and then a bird. Therefore, by considering the phenomenon of overattribution, we may seriously question the claims that the pongid subjects exhibited grammatical structure in the sequences they emitted, or even understood the representational meaning of all the symbols they often used. As mentioned, some like Terrace, Rumbaugh, and their associates, have definitely changed their original starting positions.[359]

To conclude this section, Terrace and colleagues provide an important reminder: "Sequences of symbols produced by an ape may seem grammatically related to one another in the eyes of human observers. It does not, however, follow that the chimpanzee had any knowledge of the relationships that a human observer may infer."[360]

TABLE 6

Kobo's Expletives and Insulting Signs (From Age 4 to 7)

Response	Situation
Darn Foot	One of Koko's friends whom she calls "Foot" leaves without saying goodbye
Devil rotten	Dr. Patterson asks Koko if she likes Mike, the younger male gorilla playmate
Mike nut	Dr. Patterson asks Koko if she is jealous of Mike
That rotten stink(er)	A visitor asks Koko about the identity of a photo of Koko's father
Know that stubborn donkey	Dr. Patterson asks Koko to sign concerning some drawings depicting numbers
Head stupid	Dr. Patterson puts a salt shaker on Koko's head
Rotten toilet	Dr. Patterson says to Koko, "Say bad!"
Trouble flowers	After being scolded for uprooting flowering plants

After Patterson and Linden, *The Education of Koko,* p. 150 (Table 7). These and numerous other sign combinations are considered "innovations" by Patterson by meeting a criterion for the identification of metaphor (established by Howard Gardner). Patterson writes: "Once it is established that the subject knows the literal differences between such polar adjectives as light and dark, the test simply determines whether the subject's ideas about a color being happy or sad, hard or soft, and so forth, match the collective decision of a group of adults." (Ibid., p. 147.)

Data Recording Problems: Insufficient Data Collection and Anecdotalism

Related to the problem of overattribution of certain recorded examples is the matter of the overall adequacy of the data from which researchers make their claims. Again, Terrace and collaborators,[361] including Seidenberg and Petitto, have cautioned against the use of the method of rich interpretation, especially if there is a lack of systematic data for purposes of comparison.[362] Such data should include all the signing (or other responses), plus a record of the context of these responses. The latter would include the teachers' promptings and other relevant information such as the number of choices given the ape in a testing situation, the number of trials and errors, and other pertinent information on techniques.[363]

Returning to the point of the reevaluation of their data on Nim by Terrace's group,[364] we add the following account of their explanation for the oversight of their own inadvertent contributions to the data collection:

> Painstaking transcriptions of our videotapes revealed certain aspects of Nim's signing that were not apparent to his teachers in the course of normal observation. None of Nim's teachers, nor the many expert observers who were fluent in sign language, detected either the extent to which the initiation and contents of Nim's signing were dependent upon the teacher's signing or the degree to which Nim interrupted his teachers.[365]

Thus in the last year of Project Nim, such imitations or reductions of prior teacher signing were calculated at about 40 percent of Nim's videotaped sequences; only about 10 percent were judged spontaneous, i.e., not promoted by any adjacent teacher signs.[366] They state, by way of contrast, that children at Stage I[367] have been found to imitate or reduce adjacent utterances by adults at a significantly lower percentage (18 percent), a rate that steadily decreases to a negligible amount (2 percent) by Stage V.[368]

In studying two films on Washoe—one produced by NOVA, *The First Signs of Washoe* (Time-Life Films, 1976), and the Gardners' own production, *Teaching Sign Language to Chimpanzee Washoe* (1973)—Terrace et al.[369] observed similar patterns of partial and full imitation by Washoe, with prompting by B. Gardner as well. The Gardners claim that Washoe's signing of *time eat* (as shown in the second film) was an instance of a spontaneous occurrence, yet B. Gardner had signed both *time* and *eat* immediately prior to Washoe's response.[370] Again, incomplete reporting of the context of the utterance could lead one to think that Washoe had a knowledge of time and the ability to refer to something not visible in the context (an instance of displacement; cf. chapter 7 and Conclusions on pages 146–153). Overattribution, then, may result not only from experimenter biases but also from experimenter techniques, of which inadvertent cueing may affect the animals' responses.[371] (See Experimenter Cueing, pages 139–144.)

It was already mentioned in chapter 2 that often the insufficiency in the corpus may be due to the researcher's methodological assumptions concerning the relative importance of a feature. The Gardners' early records with Washoe lacked a distinction in word order with two or more sign combinations.[372] Word order is less significant in ASL as a syntactic device than it is in English. Yet, with a pidgin variety of ASL used with the apes, it becomes the main syntactic device. What lies in the balance is whether the apes are signing creatively according to some learned syntactic rules or not. They could be using position habits or some other strategy, imitating their teacher's adjacent signing, or randomly combining signs that might be construed as "contextually appropriate."

Let's take another example in which a methodological assumption effectively leads to relaxing the criteria in reporting results. The Gardners[373] tested the ability of Washoe (and later four other, younger chimpanzees) to respond to *wh-* questions. From the results they claim that Washoe's ability would measure up to and even go beyond a Stage III child. Nevertheless they accepted *any* answer as correct so long as it was within the same lexical category.[374] Seidenberg and Petitto identify other procedural problems with the test implementation and scoring[375] to the extent that the Gardners' bold claim for Washoe appears to be quite far-fetched.

Therefore it is important to have information not only on word order but also on the criteria used to determine correct responses on tests and other testing procedures. Moreover, careful analysis of the contexts of the apes' combinations is also of paramount importance, as Terrace and associates,[376] including Seidenberg and Petitto,[377] have repeatedly stressed. Without such documentation, citation of individual utterances as evidence of some language characteristic amounts to nothing more than anecdotalism.[378]

Petitto and Seidenberg charge Patterson (especially) with a lack of reported systematic data on which to base her conclusions; i.e., relying heavily on individual examples without any available corpus containing a substantial number of transcribed utterances. Yet, they argue, without such a corpus "these examples are impossible to interpret [i.e., unambiguously]."[379]

Patterson's reliance on context for determining sign meaning, they argue, provides "a rich source of mistaken attributions since there is no evidence that Koko utilized the same contextual information in producing her signs as Patterson used in interpreting them."[380]

To show the combined effect of (1) the omission of pertinent contextual information and specific prior training, (2) a propensity to interpret mistakes metaphorically (as if "teasing"), and (3) inadvertent prompting (see Experimenter Cueing, pages 139–144), consider the account concerning Koko's "rhyming" (pp. 38–39).[381]

Note that we are not told how many toy animals were in front of Koko, what they looked like (was the pig big, the bear hairy?), what order they were arranged in, whether the teacher had trained Koko with any rhyming of individual toy animals prior to this time, or even how her teacher conveyed the sense of "rhyming" to Koko in the first place. Moreover, when Koko answered an identification question wrongly, e.g., *Pig cat*, the teacher's response was "Oh, come on," leading the reader to believe that Koko was "kidding" (and not being "innovative" at this point). So Koko was given another chance. Finally, we do not have any information about possible inadvertent cueing, like the teacher's eye movements, facial expressions, or other potential cues for Koko. To the latter phenomenon we will turn in the next section.[382]

Furthermore, without a definite corpus of sequences available, the citing of individual examples could, for all we know, be taken from an unspecified number of what might be random or otherwise anomalous gestures.[383]

A related matter concerns the ape-sign researchers' lack of presentation of the criteria used to interpret the anecdotes: "A consistent pattern of overattribution is seen throughout Patterson's paper.[384] She ascribes very specific meanings to the ape's signs without presenting any discussion of the criteria which led her to conclude that the ape intended these meanings."[385] Patterson writes, "Although Koko has produced uninterpretable strings (as do some children) most of her utterances are well suited to the situation."[386] Again we are left without any systematic distribution of data to determine her criteria for judging suitability to the situation.

A lack of reporting of sufficient data is one thing, but it is quite another to use constructs from child language acquisition studies while failing to use them appropriately. Thus, for example, all the ape-signing researchers have reported lengths of combinations as Mean Length of Utterance (MLU, see chapter 5)—as measured by signs, not morphemes—in order to make comparisons with levels of child development stages. Yet, as Terrace's group[387] rightly states, an ape's increase in MLU (e.g., with Koko up to 2.0 at age 56 months)[388] does not necessarily indicate anything about the internal complexity of the sequence. MLU was devised by psycholinguist Roger Brown to correlate with certain syntactic development as an index of a child's linguistic development, not with a mere increase in the number of words in utterances. The numerical increase in the apes' sign sequences is largely attributable to repetition and redundancy.[389]

Experimenter Cueing: The Clever Hans Effect

Another factor affecting the outcome of any experimentation with animals is the phenomenon of inadvertent experimenter cueing, or the *Clever Hans effect*. This kind of prompting was named after the famed horse at the turn of the century that seemingly could, by the tapping of its hoof, "think" by "spelling" words, "reading" sentences, and "performing" arithmetic operations. It was discovered by psychologist Oskar Pfungst, however, that Clever Hans's familiarity with its master, von Osten, enabled it to perceive as unintended signaling such imperceptible head movements as slight as one fifth of a millimeter deflection.[390] A noted scholar on animal psychology, Heini Hediger observes that "the closer—in the spatial and psychological sense—the experimenter and the animal find themselves, the greater is the danger of mutual influence."[391]

Commenting on the apes trained to use language, Hediger raises the question, "How can we prove that such answers [the chimps' productions assumed to be language-like] are to be understood as elements of a language, and that they are not only reactions to certain orders and expressions, in other words simply performances of training?"[392] He notes, "When an outside observer follows and tries to understand the signalling of a chimpanzee trained in American Sign Language (ASL), he has great difficulty in distinguishing preparative from conclusive movements [the real signs], especially since these movements succeed each other very rapidly."[393] Thus Hediger suggests that the signing apes, while having learned some ASL signs, give the appearance of knowing more than they do. The Rumbaughs concur by pointing out that "both the chimpanzee and the experimenter typically know, given the context, which subset of signs or symbols is appropriate, though the chimpanzee does not know which particular symbol within the subset is the correct one to execute. It therefore hesitantly or sloppily or repetitiously or hurriedly executes some hand movement while closely watching the experimenter's face."[394] Given a high degree of iconicity of ASL signs with nonsymbolic pointing gestures included, it is no wonder that experimenter bias, for one thing, can greatly compound the problem of determining what is or is not an instance of signing[395] (see further discussion above).

Citing evidence that even rats' test performance could vary if there is a change in experimenters,

Hediger broadens his critique of the Clever Hans effect:

> More and more I come to the conviction that the results of animal experiments do not so much depend on the exact sequence and make-up of the experiment itself, but to a great deal also on the past of the individual animal and on the personal attitude of the experimenter. These two important facts are very often what conventionally is neglected most.[396]

Thus the Sebeoks note that because any animal experimentation requires direct or indirect human contact, some subjectivity on the part of the experimenter must result. This is *especially* so for apes "which can scarcely survive in the absence of an intense level of social intercourse with their handler before they are capable of undergoing the rigors of training for any task."[397]

Patterson, in her most recent writing (*The Education of Koko*), claims that in contrast to other experimenters like Terrace and Premack, she has developed a close rapport with her ape subject. In fact, she observes, "it is hard not to wonder whether the different conclusions about ape language abilities reached by these scientists ultimately trace back to the different relationships between experimenters and subjects and to the persistence that has marked the efforts of those of us who have established close rapport with our subjects."[398]

Isn't this frank statement that Koko performed better for Patterson than, say, Nim did for Terrace, readily an admission that what is being studied is a function of a clever animal and a desire on the experimenter's part to assign language ability? If real language ability were within the capability of apes, why wouldn't *all* the experimenters agree on its manifestation? Why, after all the carefully devised experiments to induce the apes to learn, do the researchers disagree so much among themselves? Along these lines, Chomsky has often commented, "It is difficult to imagine that some other species, say the chimpanzee, has the capacity for language but has never thought to put it to use."[399]

Commenting perceptively on the state of the wide-scale criticisms among the various ape researchers, ethologist Donald Griffin observed a few years ago:

> The leading groups of experimenters training apes to use symbolic communication share a general faith in the significance of the endeavor, but they are often critical of each other's specific experimental procedures. If one accepts the sharpest of these mutual criticisms, it is tempting to dismiss all of the results, as inconclusive.[400]

It should not be surprising, then, that the Clever Hans effect can creep into supposedly objective teaching procedures that attempt to rule out the experimenter cueing (as with Premack's plastic token procedures for Sarah and Rumbaugh's simulated-language machine for Lana).

In an attempt to eliminate such prompting in the ape-signing projects, the Gardners[401] and, to a less extent, Patterson[402] have employed a "double-blind" testing procedure (see Figure 10, page 143). The double-blind test purports to test an ape's knowledge of vocabulary, roughly as follows: The ape is shown an object (or a picture of one) randomly selected by one experimenter, whom the subject cannot see, from a stock of items that the ape has learned to sign; a second experimenter, who cannot see the object, records the sign made by the ape.[403] The Gardners report a far greater than chance degree of accuracy (71 percent),[404] while Patterson reports a 60 percent degree.[405]

Do such test results indicate that the signing apes have indeed learned to label items correctly? Not necessarily. Both Seidenberg and Petitto[406] and the Sebeoks[407] have put forth a number of potential cueing situations in the absence of specific information on the method of presentation of the items (order and number of times), manner of blocking the trials, prior training for the test, and other performance on multiple tests. The Sebeoks are especially meticulous in their analysis of possible experimenter effect.

Commenting generally about the administration of double-blind tests with ape subjects, the Sebeoks note the following:

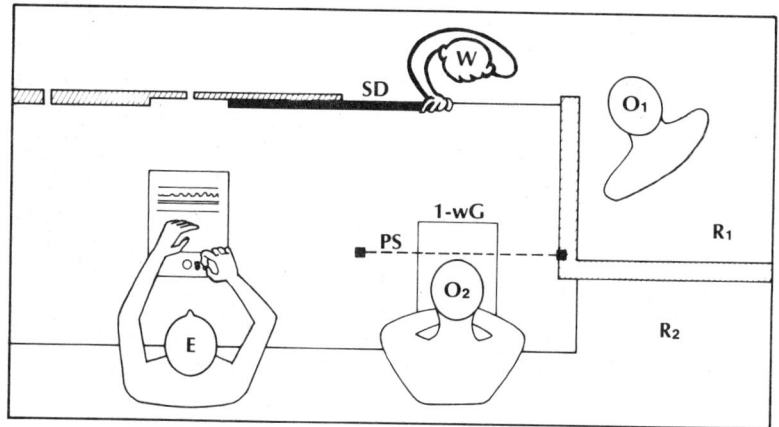

FIGURE 10. Double-blind Testing Apparatus Used in Project Washoe.

R_1 R_2—two rooms of the laboratory
PS—projection screen, for back projection
W—Washoe
O_1—First observer, out of auditory and visual contact with O_2
O_2—second observer, unseen by Washoe, and unable to see screen and experimenter, confirms observations of O_1
1-wG—one-way grid, or screen
SD—sliding door for Washoe to signal for show to begin
E—experimenter, operating slide projector

Source: Adapted from R. Allen Gardner and Beatrice T. Gardner, "Comparative Psychology and Language Acquisition," in T. A. Sebeok and J. Umiker-Sebeok, eds., Speaking of Apes: A Critical Anthology of Two-Way Communication with Man (New York: Plenum, 1980), pp. 318–19.

Researchers have uniformly found it more difficult to test their ape subjects under the more stringent conditions of a double-blind situation than in normal, more sociable circumstances, and both chimpanzees and gorillas alike, whether using ASL or an artificial language, have shown a marked deterioration in their performance in the former, more artificial type of setting. The lack of cooperation on the part of an ape in such situations is usually described in reports in general terms rather than in detail, with frequent allusions to the animal's psychological or emotional attitude toward the "blind" experimenter.[408]

As an example of this difficulty, they mention Patterson's remark of Koko's "avoiding the task" by various means such as responding to all the objects with the same sign or refusing to sign at all.[409] The

Sebeoks doubt whether the 60 percent accuracy figure included the avoidance measures as errors.[410]

We are left, then, wondering whether there are any fool-proof measures to determine the knowledge of true symbolism in the apes. As mentioned earlier (pp. 132–133), the Rumbaughs simply think there is a lack of theoretical criteria in the literature to distinguish between a conditioned response and a truly symbolic one in these animal experiments.[411]

Further support for the apes' cleverness in picking up cues from experimenters comes from a study by primatologist Emil Menzel.[412] A group of young chimpanzees were placed in a field where various treasures were hidden. Only the leader chimpanzee was shown the various objects, which included toys, various amounts of food, and some fake snakes (snakes frighten chimps). The leader was then placed among the group and left to communicate the information. The previously "uninformed" chimps typically ran ahead of the leader to secure the treasures and responded in such ways as to demonstrate their knowledge of the type of treasure they were seeking. Menzel observes, "In sum, the chimpanzees were able, and without any deliberate training on our part to convey to each other the presence, direction, probable location, and the relative desirability and undesirability (if not the more precise nature) of a distant, hidden goal which no one had directly seen for himself."[413] Thus Menzel concludes that the ASL trained chimpanzees cannot communicate more information to each other than what others in the wild can do with their eyes.[414]

Problem Solving, Not Language Learning

We have already referred to the problem of the reward-reinforcement conditioning basis of so much of the ape language experimentation. We have seen this in the shaping of responses with Sarah and Lana and in the heavy use of forced-choice tests in "teaching" vocabulary (pp. 130–133).

We will now consider various critiques of D. Premack's language-board experiments and Rumbaugh's computer-language board. Terrace[415] has analyzed

Premack's experiments that are described in his later work, *Intelligence in Apes and Man*,[416] incorporating the 1972 study previously cited. Applying the law of parsimony to the various claims of Sarah's linguistic ability, Terrace found Premack's attribution of linguistic meaning to the plastic chip arrangements to be unwarranted. Routines similar to what pigeons can learn in pecking experiments could adequately account for Sarah's behavior:[417] Pigeons were trained to peck the sequence A→B→C→D, where A,B,C, and D were different colors, at levels of accuracy comparable to that reported by Premack in the case of "four-word sentences" (e.g., "Mary give apple Sarah").

The symbols could just as easily have been nonsense words, since there was no real language usage involved. That is, there was no contrast with other symbols, essentially no other variation, except the particular object in question. Regarding the recognition of the trainer's name, all Sarah had to do was match the symbol that the trainer wore around the neck with the symbol on the board as the first of a rotely learned sequence.[418] Thus, Terrace argues that the general form ("Trainer give X recipient") "could just as well [i.e., more parsimoniously] be described as rotely-learned sequence ABXD, in which substitutions of the meaning of X varied with the object on hand."[419] Terrace suggests that such problems might also be within the capability of a pigeon.[420]

How about Sarah's comprehension of "Sarah put apple pail banana dish" as indicative of a knowledge of hierarchical sentence structure (see chapter 2)? Terrace observes that since Sarah had only one kind of instruction (insert or take), it is not surprising that she performed the same action twice (both objects were *outside* the receptacles in the first place). By the mere location of the symbols in the string, Sarah would figure out what was supposed to go where.[421]

Consequently Terrace concludes, "The homogeneous nature of the questions posed during any one session, along with the restricted range of possible answers increases the likelihood that nonlinguistic contextual clues contributed to the performance of

Evaluation of Ape Communication with Man

Premack's subjects [Sarah and other, younger chimps]."[422] In short, Sarah solved problems for rewards.

By the end of the 1970s, Premack had changed his own mind about the claims he had earlier made, stating that "language training does not instill human language [in chimps].*[423]

Similar criticisms of rote-type learning have been leveled against the Rumbaughs' earlier work with Lana,[424] as the Rumbaughs essentially recognized later.[425]

In fact, psychologists Thompson and Church present evidence that Lana's behavior "can be simulated by a computer model in which the animal selects, depending upon context, one of six stock sentences with fixed and variable elements."[426] They demonstrate that on the basis of yes or no answers to these questions (Object present? Object food? Object in machine?), a stock sentence was chosen. They therefore attribute Lana's behavior to paired-associate learning (i.e., learning lexigrams for objects, events, and people) and conditional discrimination learning (i.e., given a particular context, learning to produce a stock sentence like, "please machine give [INCENTIVE]" with variable lexigram substitution for the incentive slot).[427]

Although Lana's behavior may be considered more complex than Sarah's, the above analyses point to the fact that these chimps were engaged in various problem-solving routines, not real language learning. More generally we see again that simpler explanations, which don't assume linguistic ability, are adequate to account for the signaling behaviors of the apes discussed in this chapter.

Conclusions

Having gone over most of the major criticisms of linguists, psycholinguists, and comparative psychologists concerning the various ape language projects,

---

*Premack thought such training did make "some overall cognitive changes," however.

in this section we will review the twelve characteristics of human language in order to see their applicability to the apes' behavior.

It is important to emphasize, though, that the critical literature has substantially increased to the extent that the aforementioned works are representative, not exhaustive, of the mounting skepticism, not only by traditional "opponents"—those who have held that language acquisition is species-specific—but also by psychologists who have been directly involved in the ape-man communication research projects.

With respect to the following evaluation, it must be said that every one of the characteristics claimed by certain psychologists to be within an ape's capability (and not all of them have been so claimed) has been seriously questioned on the basis of the previously discussed objections. These, then, are the distinguishing features of human language that we have identified in chapter 7.

1. Discreteness of Sound/Form

As we have seen with the Hayeses' attempts at teaching Viki speech sounds, the inability of a primate to phonate is not even remediable. Philip Lieberman[428] has made it clear that the chimpanzee does not have the same upper-vocal-tract configuration necessary to produce unaided the most fundamental vowels (*i, u,* and *a*).

With respect to the manifestation of discreteness in sign language, recall that various psychologists[429] have called into question a number of matters concerning apes' signing behavior, including random movements that have at times been interpreted as "signs."

2. Symbolic Nature of the Units

The above-mentioned researchers have made the point that there is little evidence that the apes have the same kind of knowledge of the representational nature of their signs or manipulable symbols. "They [the signing apes] have learned that the mere behavior of producing signs could be used to effect certain outcomes, e.g., getting food, social approval, release from work time, and the like. This behavior is similar to that of a very young child who has learned to say some things that it does not understand."[430]

As mentioned earlier (p. 130), the Rumbaughs criticize the first ape-language investigators for assuming they could simply employ notions from child language research without question.[431] They single out two presumptions regarding the symbolic nature of words: "(1) for a chimpanzee to learn language, it must first learn which members of a set of arbitrary symbols are to be associated with which real-world events and objects; (2) as the chimpanzee learns these associations, he will bring to the task an inherent capacity for representational and symbolic communicative ability that will enable him to use these symbols in a variety of ways."[432] As we have seen, however, on the basis of their own work, the Rumbaughs have concluded that "representation or symbolization is not inherent in the chimpanzee's capacity to select a symbol when presented with an object, action, or state." Yet while arguing that the ability to label objects "is not a sufficient criterion for the symbol to be viewed as a word," they maintain that symbolic representation "can be acquired by the chimpanzee."[433]

Following Chomsky, let us assume that the determination of word meaning "appears to involve other systems of knowledge and belief in an intimate and perhaps inseparable way."[434] Then we may regard an ape's "labeling" ability to be indicative of its general level of intelligence. After all, other animals, no less than the great apes, have to be credited with the ability to recognize classes of objects. For an ape to use objects of the same class (through associative learning and reinforcement) ought not to be surprising, then. How truly symbolic this labeling is, is still open to question, though a limited amount appears possible (as with the Rumbaughs' Sherman and Austin). Again, we stress that much of the discussion concerning the ape's knowledge of symbols is tainted with the researcher's *intention* that the signs mean such-and-such.[435]

To conclude, we quote Adrian Desmond's reaction

to Patterson's numerous glosses of Koko's "abstract" vocabulary: "What on earth does a gorilla understand by 'funny', and how exactly could its *human* meaning have been conveyed by a symbol? This is the crux; if an ASL-derived sign for 'funny' is listed in Koko's lexicon, we are lured into the assumption that she understands its English meaning."[436]

3. Rule-governed system

Perhaps the strongest of the claims made by the Gardners and others have been met with the strongest of the criticisms, especially by Terrace and his associates: "In sum, evidence that apes can create sentences can, in each case, be explained by reference to simpler nonlinguistic processes."[437] We have seen that the sequences of signs can be interpreted as "sentences" only by an unjustified use of the method of rich interpretation. As Chomsky has emphasized, we can correctly say that "children are exhibiting 'incipient human language behavior' only because of the later states achieved";[438] but not so with apes. In fact, Terrace's group is careful to point out that neither Nim's nor Koko's MLU increased beyond a certain point (1.6 and 2.0 signs per utterance, respectively),[439] to say nothing about complexity. As mentioned in chapter 2, pages 32–34, Terrace pointed out Nim's very repetitive style in longer sequences.[440]

Without strong evidence for *structured* sequences of signs, we certainly cannot say that chimpanzees have learned a system of rules. Whatever regularities in word order Terrace and his colleagues found in Nim's sequences were attributed to partial or complete imitations of the teacher's adjacent utterances and to lexical regularities in the cases of two-sign combinations;[441] these and longer sign combinations proved to be highly repetitive. Seidenberg and Petitto suggest, as an alternative, that the apes could be following certain strategies in order of sequencing: "sign from the pool of favorites" (in the case of Nim), "sign until the experimenter terminates the trial," or "randomly select signs from the vocabulary of responses until you hit the correct one."[442] Strategies to gain rewards, suffice it to say, are *not* the same as the abstract rules of grammar that make up the system of knowledge that a child does acquire.

4. Compositional

Where human language exhibits the influence of abstract phrase structure in the determination of the meaning of a sentence, clear data (i.e., without evident overattribution) on the apes in this regard are definitely lacking. Recall that the Premacks' attempt to show the compositional nature of Sarah's "comprehension" of an elliptical string proved fallacious because a simpler, nonlinguistic explanation could easily account for the phenomenon. The Sebeoks refer to Lenneberg's (1975) experiment, using human subjects, in an attempt to replicate the Premacks' work.[443] The high school students "quickly outperformed Premacks' chimpanzees, [but] they could not correctly translate into English any of the sentences which they themselves had completed. Thinking they were merely solving puzzles, the subjects failed to connect the plastic tokens with language."[444]

5. Complex

This feature of language has not, to our knowledge, even been directly addressed in the ape language literature. The complexity of a sentence demonstrates the highly abstract nature of the form and meaning relation. Akmajian and colleagues quite rightly ask what it would mean for the apes to comprehend a structurally ambiguous sentence: Could they group the words (chips, etc.) differently in such a way that distinct phrasal organizations underlying the string would have to be postulated in order to account for their knowledge of the ambiguity?[445]

It seems inconceivable—barring flagrant overattribution—that an ape could produce (in sign or simulated language, of course) a structurally ambiguous sentence along the lines of "I enjoy boring researchers," or a counter-factual conditional proposition analogous to "I would have eaten the bananas you gave me earlier if I hadn't been so frustrated."*

6. Displacing

Patterson and the Gardners have attempted to demonstrate this characteristic in their respective subjects. Patterson, for example, refers to Koko's use of "time" expressions like "yesterday" and "tomorrow"—forms that are made with reference to the "time line" extending from behind the body to in front of it.[447] While Patterson claims that Koko has a sense of past and future time,[448] the use of a word like *tomorrow* could very well be an avoidance means to postpone something undesirable. Yet we know that these animals are endowed with great memory capacities and thus we need not dismiss all the uses of time words as clever tricks to fool the researchers! However, what a far cry these supposed examples of time displacement are from the notion of displacement in human language, in which one can express abstract, hypothetical, and imaginary thought.

Animal psychologist Heini Hediger flatly states, "To my knowledge, up to now, no animal, not even an ape, has ever been able to talk about a past or future event."[449]

7. Unbounded in Scope (Open-ended)

The creative aspect of language use is in sharp contrast to the highly imitative and often repetitious signing or symbol manipulation of the apes. Claims concerning their creative signing have been adequately answered in the literature drawn upon in the previous sections. Moreover, the scope of the apes' sequences have largely, in some cases almost entirely, been limited to the request for some reward.

8. Independent of Stimulus Control

There is little that can be said for the apes' signing or visual symbol usage that does not entail the care-

---

*In ASL, simultaneous modulation of a sign (modification by the nondominant hand) is used to convey lexical ambiguity.[446]

fully controlled conditioning and/or prompting of responses elicited by the experimenter or experimental conditions. With respect to a discourse analysis of Nim's signing, revealing his frequent imitations and interruptions of his teachers, Terrace and his coworkers have reported that "Nim's use of language differed *fundamentally* from that of a child" (italics added).[450]

9. Suitable for Contextualized Communication

This particular feature might appear to be shared by the apes. However, as discussed earlier, the fact that the apes draw from a comparatively small stock of lexical items, mostly related to the meeting of needs, tends to be interpreted as "suitable" on nearly every occasion.[451] Koko's greater stock of reported signs, it must be remembered, does not indicate so much a wider range of conversational topics as it does a far greater extent of the researcher's *intentions* for the subject.

10. Independent of Need Satisfaction

This point hardly needs consideration. Terrace and his colleagues have emphasized that "the function of the symbols of an ape's vocabulary appears to be not so much to identify things or to convey information . . . as it is to satisfy a demand that it use that symbol in order to obtain some reward."[452] (While Patterson's examples of Koko's exchanges appear less restrictive than, say, Nim's, there is generally no indication from Patterson that Koko's "successful" responses went unrewarded.)

11. Spontaneously Acquired

There is absolutely no evidence that the apes would ever have learned anything of the sort—not to mention what they have learned—without the direct and systematic intervention of humans. (In contrast, one has to intervene to keep a child from talking!) Children, as has already been observed in chapters 5 and

6, need no special instruction or shaping to learn a language; mere exposure to its use suffices to trigger the innate mental structures (the faculty of language) into guiding them to acquire the correct set of rules underlying the limited data.

Evaluation of Ape Communication with Man

12. Culturally Transmitted

As noted in chapter 2, there have been reports of second-generation chimps learning some signs from the original subjects. Being highly social animals in the wild, the great apes would be expected to influence one another in their colonies in captivity. Hence, the passing on of signs by the trained apes could almost be assumed, particularly because of the desirable consequences for the learners with their human captors.

So Fouts's report that Loulis has learned signs from Washoe should not surprise us: if the behavior works for Washoe, why shouldn't she pass on the "generalized motor patterns" to other chimps?[453] Future studies concerning the long-term effect on the natural communication system of these apes will obviously be of interest.

A final word from Herbert Terrace, quite ironically, sums up the still-unexplained gap between human languages and any natural or conditioned animal communication system:

> Despite the frustrations of Project Nim, I knew that there could be no substitute for that intelligent bundle of playfulness and mischief, a creature more human than any other nonhuman I knew. One of the reasons this parting was so painful was that *there was no way to talk with him about it.* Nim and I were able to sign about simple events in his world and mine. But how could I explain why I and the other project members who came along to Oklahoma suddenly abandoned him? How could we explain that it was necessary to leave him forever in a totally new environment, with a totally new group of human and nonhuman primates? (italics added).[454]

A Sobering Postscript From an Unexpected Source

Psychologist David Premack expressed, in the early 1970s, some of the strongest claims about the ability of his subject (Sarah) to learn and use a language-like system. However, in 1978, as personal communication to Adrian Desmond, Premack voiced a clear change of position:

> Chimps do not have any significant degree of human language and when, in two to five years, this fact becomes properly disseminated, it will be of interest to ask, Why were we so easily duped by the claim that they do?[455]

chapter nine

Accounting for the Language Gap

In the previous chapters the reader has been presented evidence that there is a definable gap between human linguistic ability and primate communication. Explanations of the language gap have come from numerous sources, especially evolutionary theory. The authors report on these hypotheses, then offer their own answer to the problem.

On the basis of the foregoing analysis of the chimpanzee and gorilla language projects, we have seen that there is no sound evidence to refute the traditional view that language acquisition is species-specific. If human language were a product of general intelligence, then would we not expect to find such intelligent animals as the great apes learning the rudiments of language, especially after all the extensive methodical attempts, lavished with plenteous rewards?

We have seen that an innate (genetically determined) mental schemata—the language faculty, with its universal principles of language—is necessary to postulate in view of the facts of language acquisition, the rule-governed properties of human languages, and the nontrivial similarities among languages at a highly abstract level. Thus we may conclude that such biological endowment responsible for the acquisition of a generative grammar (upon exposure to language) is *not available* to these very intelligent primates.

Chomsky reasons that while the apes may share with humans certain aspects of what he calls "conceptual structure," they lack "computational structure" (i.e., the human language faculty).[456] Some sharing in the conceptual structures permits the apes to make use of signs and other indicators, to label objects in a limited fashion, while the encoding thereof in terms of a structured system is lacking. In short, humans are the sole possessors of a "Language Acquisition System" (LAS).* Chomsky comments that if we were to find a faculty in other organisms like the human language faculty, it would be "a kind of biological miracle" because of the "enormous selectional advantage" of having such a system: "It is difficult to imagine that some other species, say the chimpanzee, has the capacity for language but has never thought to put it to use. Nor is there any evidence that this biological miracle has occurred."[458]

The Framework for Debate: Why a Language Gap?

We thus turn to the matter of framing some account of this language gap by surveying and critiquing several approaches within a Darwinian or Neo-Darwinian framework and finally offering another alternative—a creationist view.

Darwin did not offer a specific account of the beginnings of human language in *The Origin of Species* (1859), but he did treat it in *The Descent of Man* (1871) in the context of a comparison of man's mental powers and those of other species. In his treatment of the various historical answers to the origin-of-lan-

---

*Or, LAD, Language Acquisition Device, so called by Chomsky.[457]

guage question, James Stam states: "It was his [Darwin's] purpose, here as elsewhere, to show the genealogical links among species, without obliterating specific differences."[459] Darwin concluded, "The lower animals differ from man solely in his almost infinitely larger power of associating together the most diversified sounds and ideas."[460]

In *The Expression of Emotion in Man and Animals* (1872) Darwin, according to Philip Lieberman, links man's emotive expressions (vocal and gestural) with animal behavior, but does not deal with the syntactic organization of either human language or animal communication.[461] Thus Lieberman thinks that Darwin failed to apply to his discussion of language his general principles stated in *The Origin of Species*, namely, natural selection and its corollary, incremental development.[462] Darwin "does not really attempt, through evolutionary process, to connect human language with the communications of other languages."[463] Moreover, Darwin's "basic underlying premise [was] that human language is disjoint from the expression of emotion."[464]

Nevertheless, more important was the legacy of the Darwinian framework for future expositors, namely the theoretical mechanisms of *natural selection* (and related principle of *preadaptation*—"natural selection channeling development in a new direction because of previous modifications for some other role")[465] and *mutations* (random changes in the genetic structure of organisms).

Explanations of the Gap Within Evolutionary Theory

Underlying the current debate concerning the species specificity of human language is the concern of certain comparative psychologists to provide evidence for the commonly held *continuity* hypothesis of the evolutionary development of language—a link between human language and animal systems, including the recent pseudo-language of the apes. This view can, in some measure, be seen as a more modern, sophisticated replacement of older naturalist views of human language developing from imitative cries of nature, emotional responses of early man, etc., as, for example, expressed by Rousseau in the eighteenth century.[466]

On the other hand, the *discontinuity* hypothesis draws heavily on the mechanism of mutations to account for the *emergence* of a qualitatively different kind of mental structure, the language faculty. A third evolutionary approach stems from Lieberman's reliance on preadaption as an attempt to incorporate elements of each of the two other hypotheses; it assumes that many small quantitive changes lead to an overall qualitative difference.[467]

We now turn to a brief treatment of each of these views, in light of what is currently known about human language.

The Continuity Hypothesis

Much of the impetus of the recent ape-human communication experiments has come from the concern as to whether man's supposedly closest living relatives, the great apes (especially chimpanzees) could exhibit some languagelike behavior. Implicit, then, is the notion that there must be a continuity in the evolutionary development of man from the early hominids. Francine Patterson is candid about her view and its significance in her research with Koko the gorilla. Commenting on the impact of Darwinian thinking, she says:

> For all the changes in the theory of evolution, it supplies a perspective that permits the search for significant continuities between man and animal. It is difficult to overstate how different the same behavior might seem when seen from an evolutionary perspective as opposed to the traditional animal/human dichotomy.[468]

Such a commitment to a continuity version of evolutionary theory can be seen in Patterson's strong claims and liberal interpretations of Koko's performance, as we have already seen in chapter 8.

For their part, the Gardners emphasize the importance of finding general "laws of behavior" common to all forms of organisms and their behavior, whether "simpler" or "more complex."[469] While not speculating about the origin of human language from simpler to more complex forms, the Gardners stress "general functions" in relating the various kinds of behavior.

Commitment to a continuity evolutionary hypothesis, then, would seem to prompt one to look for incipient language behavior in the pongids, in order to disclaim man's linguistic uniqueness. Furthermore, such a commitment would imply the need for a mechanism common to all organisms to account for the new behaviors; hence, the importance of "learning," as stressed in *empiricist* theories of epistemology. Recall that Chomsky strongly attacked the logic underlying this conclusion.[470]

We now come to an analysis of the continuity hypothesis of the origin of human language—i.e., its development from more primitive systems, as presented by philosopher Karl Popper and reviewed by Noam Chomsky in *Language and Mind*.[471]

In his "Clouds and Clocks" lecture, Popper espouses the view that the present "higher stage" of language has evolved from a "lower stage" that could only express emotional states.[472] Chomsky criticizes this continuity view because Popper "establishes no relation between the lower and higher stages and does not suggest a mechanism whereby transition can take place from one stage to the next. In short, he gives no argument to show that the stages belong to a single evolutionary process."[473] Chomsky further observes, "In fact, it is difficult to see what links these stages at all (except for the metaphorical use of the term 'language')."[474]

In other words, argues Chomsky, there is a definite gap between a vocal expression of emotion and an instrument of abstract thought:*

---

*This criticism would thus hold for another naturalist view—the "onomatopoeic" view of the origin of human language, i.e, that it evolved from early man's attempts to imitate the sounds of nature. It does not take much to refute this view: All we need to do is ask speakers of different languages to say how a dog or a chicken sounds in their languages. Each language apparently has its own version of such an "imitative" sound. We must conclude that the sound-meaning relation in a language is arbitrary or conventional (truly *symbolic*, as discussed in chapter 1).

There is no reason to suppose that the "gaps" are bridgeable. There is no more of a basis for assuming an evolutionary development of "higher" from "lower" stages, in this case, than there is for assuming an evolutionary development from breathing to walking; the stages have no significant analogy, it appears, and seem to involve entirely different processes and principles.[475]

Indirect disconfirmation of the continuity hypothesis comes from a lack of any mammalian species except man to make speech sounds; instead, surprisingly, certain bird species can imitate these sounds. Chomsky cites ethologist W. H. Thorpe's comment that such birds therefore "ought to have been able to evolve language in the true sense, and not the mammals."[476]

Thorpe, however, attempts to demonstrate a relationship between animal communication systems and human language with respect to three supposedly shared characteristics: "purposive," "syntactic," and "propositional."[477] In each case, Chomsky argues that when these terms are applied to animal systems, the level of abstraction is such that any kind of behavior may be so described.[478]

We turn full circle, then, to the point made in chapter 7: human language is unique as a system of knowledge of sounds and meanings, without analogue among animal systems of communication. Further, man's putative closest living relatives, the higher apes, have nothing comparable, nor have any been taught the rudiments of a human language system, despite the efforts and desires of researchers. And while attempts to train apes to produce speech sounds have been a failure, parrots and mynah birds have an amazing ability to mimic many human speech sounds. We should reiterate at this point that the expectation that primitive forms of language should be found among so-called primitive peoples has also been resoundingly disconfirmed.[479]

In view of the fact that human language "is based on entirely different principles" from those in animal systems, Chomsky's unequivocal position, then, is that "it is quite senseless to raise the problem of explaining the evolution of human language from more primitive systems of communication that appear at lower levels of intellectual capacity."[480]

Thus this unique "mental organ" of the language faculty, without even an *embryonic form* appearing in an otherwise intelligent ape—so far as can presently be determined—stands as an anomaly in the aforementioned continuity hypothesis.

The Discontinuity, or Emergence, Hypothesis

While most of the psychologists making the strong claims about their ape subjects (see pages 121-23) predictably accept a continuity model of an evolutionary account of language development, for his part Chomsky disclaims any necessary commitment to a "continuity doctrine" within evolutionary biology. He adds, though, that "the initial commitment is an entirely reasonable one, but some care must be taken in drawing specific consequences from it."[481]

Within an overall evolutionary scheme, then, a continuity model of language phylogeny (development) can be said to be "reasonable"—perhaps more so than a discontinuity model. Nevertheless, there is nothing comparable to human language in the animal kingdom. If the apes could be shown to learn how to form bona fide grammatical strings, if there were animal systems of communication that were even somewhat similar to human language, if there were records of simpler human languages, etc., then it would be much easier to justify the continuity hypothesis than it actually is.

Chomsky admits that this gap "poses a problem for the biologist, since, if true, it is an example of true 'emergence'—the appearance of a qualitatively different phenomenon at a specific stage of complexity of organization."[482] Concerning this evidently radical change in mental organization, Chomsky elsewhere remarks that "it seems reasonable to assume that evolution of the language faculty was a development specific to the human species long after it separated from other primates."[483]

Chomsky's evolutionary account postulates some change(s) in the genetic code causing significant neural restructuring of the brain of early man, resulting in the emergence of the language faculty. Concerning just what this mutation or series of mutations must have been like, Chomsky does not have much to say. However, at one point in a discussion (in the 1975 debate with Jean Piaget) on the central nervous system, Chomsky hypothesizes that, given the already high level of complexity of early man as an organism, the addition of "a minute element" of information to this genetic code would suffice "to account for language-specific structures."[484]

Yet, as far as current knowledge can go, such mutations are, in Chomsky's notion, "biologically unexplained," though *not* "biologically inexplicable," as Piaget holds.[485] The latter considers it implausible to assume that linguistic structure is due to "a random origin" and thus attributes the origin of language to early man's self-adjusting interaction with the environment (*auto-regulation*).[486] To the extent that this hypothesis equates language phylogeny (language development within the species) with language ontogeny (individual language development), the criticism of Piaget in Fodor's fixation of belief argument still holds (see chapter 6, "Criticism of Piaget's Theory" [pp. 104–7]).

To conclude with Chomsky's account, with its heavy reliance on random mutations, Chomsky admittedly expresses a tenuousness about both linguistic and other biological developments in man:

> We can, *post hoc*, offer an account as to how this development might have taken place but we cannot provide a theory to select the actual line of development, rejecting others that appear to be no less consistent with the principles that have been advanced concerning the evolution of organisms. Although it is quite true that we have no idea how or why random mutations have endowed humans with the specific capacity to learn human language, it is also true that we have no better idea how or why random mutations have led to the development of the particular structures of the mammalian eye or the cerebral cortex. We do not therefore conclude that the basic nature of these structures in the mature individual is determined through interaction with the environment. . . .[487]

Expressed differently, "Little is known concerning evolutionary development, *but from ignorance, it is impossible to draw any conclusions*" (italics added).⁴⁸⁸

A Hybrid Hypothesis: From Quantitative to Qualitative Differences

Standing somewhere between these two proposals is Philip Lieberman's view that numerous cognitive differences between man and ape can result in an overall difference *in kind* of mental organization: "The inherent cognitive abilities of humans and chimpanzees thus could be quantitative and still have qualitative behavioral consequences."⁴⁸⁹ Basing his hypothesis on the principle of *preadaption,* Lieberman places emphasis on the step-by-step workings of natural selection to "effect radical changes of behavior."⁴⁹⁰ He cites as an example the complex cognitive patterns associated with toolmaking, because of their selectional advantage, as being transferred to human language.⁴⁹¹

His view of great changes through increment is reflected in his echoing of Darwin's point: "Evolution proceeds in small steps."⁴⁹² But Lieberman also adds the following qualification: "The only reason that human language appears to be so disjoint from animal communication systems is that *the hominids who possessed 'intermediate' languages are all dead.*"⁴⁹³ The matter, then, boils down to postulating either a long series of quantitative neurological changes and physiological changes in the vocal tract (which add up to an appreciable qualitative difference) or a great change or series of changes at one stage of early man's development, thereby setting him apart linguistically from the other primates. Lieberman focuses on the development of the upper vocal tract as a precondition for the origin of human speech:

> No other living animal appears to have a language that makes use of encoded acoustical signals. Neural mechanisms appear to exist in *Homo sapiens* that are involved in the decoding of speech signals. These neural mechanisms would appear to be the result of a long process of mutation and selective adaptive pressures. *Homo sapiens* also appears to have a "unique" supralaryngeal vocal tract that has adapted to the particular re-

quirements of encoded speech. The supralaryngeal vocal tract of *Homo sapiens* is essentially "matched" to the neural mechanisms that decode speech.[494]

In reviewing a slightly earlier version of Lieberman's theory, linguist David Crystal criticizes Lieberman for overstating his case in view of the speculative nature of the description (such as given in the preceding quotation).[495] Besides the problem of dogmatism, Crystal points out two further shortcomings that are instructive for us at this juncture: (1) Lieberman forms a conclusion that does not follow from the evidence provided and (2) he makes an unwarranted attack on other theories of language origin, particularly the special creation view. For the first, Crystal mentions that although Lieberman *disavows* the uniqueness of human language, if anything, *he provides substantial evidence for that uniqueness*. As we have seen in Lieberman's later work (1975), he postulates qualitative differences from numerous quantitative ones. He also employs a missing-link argument to account for the disparity between the communication systems of the great apes and human language.

With respect to the second objection, Crystal observes:

> At best he [Lieberman] has only provided a persuasive argument for taking an (N.B., not THE) evolutionary hypothesis seriously. It goes too far to say the "modern man's speech-producing mechanism has CLEARLY evolved, through the Darwinian process of mutation and natural selection, from an ancestral form that is similar to the vocal apparatus of living non-human primates" ([p.]3). . . . This he has not shown, nor can show, at present. He himself admits that we know nothing about the "bridge" that has to be built between the older fossil hominids and modern man ([p.] 37), and it is difficult to see how one can in principle get at such information as the phylogenetic nature of rapid articulatory movement or the relationship between skeletal structure and soft tissue, which are essential to any argument (emphases his).[496]

It is important to emphasize, therefore, that at this level of speculation no empirical distinction may be made between a discontinuity hypothesis such as Chomsky's or a modified continuity view such as

Lieberman's, as Crystal's comment would seem to imply.

Where does this state of affairs leave us? If, as Chomsky has effectively argued, human language is a product of a certain type of mental organization, not shared by any other species, then how can an evolutionary view of man's biological development encompass the species-specificity of human language acquisition?

As we have already seen, to account for any evolutionary development there must be postulated some major—albeit chance—changes in early man's central nervous system and vocal-tract configuration. These changes, then, together with what would have to be already complex cognitive structures, would give rise to the emergence of the human language faculty. It is important to realize the consequences of a linguist's position on the uniqueness of human language. Chomsky has stated elsewhere that to invoke "evolution" as an explanation of this unbridgeable gap is at present an essentially vacuous argument:

> The process by which the human mind has achieved its present state of complexity and its particular form of innate organization are a complete mystery, as much of a mystery as the analogous questions that can be asked about the processes leading to the physical and mental organization of any other complex organism. It is perfectly safe to attribute this to evolution, so long as we bear in mind *that there is no substance to this assertion—it amounts to nothing more than the belief that there is surely some naturalistic explanation for these phenomena* (italics added).[497]

Thus the postulation of a sudden and drastic change—an emergence—in the mental organization of man's early forebears appears at present to be the most plausible way in which an evolutionary view may still be held in the face of the facts of human language.

An Alternative Hypothesis Based on Different Assumptions

What appears reasonable about a discontinuity hypothesis, relying heavily on random mutations, is its compatibility with the above-mentioned fact of the species specificity of the language faculty. Its reasonableness as an *evolutionary* theory, however, does not seem evident to everyone, especially to those who hold to the continuity hypothesis.

Francine Patterson, apparently uneasy about the attacks on her interpretations of Koko's behavior, states, "Behind Noam Chomsky's theory that there is an inherited deep structure of language,* one can see a creationist view of the universe."[498] It would appear that Patterson's commitment to behaviorist psychology (empiricism) does not allow her to view man as having qualitatively different mental structures (at least in regard to the language faculty) from those of apes.†

Is it not both curious and instructive, then, that a creationist view could be mistakenly seen in a generative grammarian's view of how human language originated?

We submit that a transformational linguist's innatism has significant features consonant with a *creationist* view in terms of (1) a distinction of man from primates with respect to (at least) one qualitatively different mental organ and, consequently, (2) its consistency as an account of man's unique possession of an instrument for the expression of abstract thought.

A Key: A Glance Back at Rationalism and Its Roots

This conclusion should not be surprising in light of the roots of seventeenth-century rationalism, from which transformational theory derives its mentalism. Descartes's dualism—the distinction between a nonmaterial soul (or mind) and physical bodies—ostensi-

---

*We should add here that Patterson mistakenly confuses the notion of deep structure—which is part of the theory of generative grammar—with that of the language faculty with its universal principles of language (universal grammar) constraining the forms of natural languages. Cf. Chomsky, *Language and Responsibility*, p. 171, concerning the confusion between the notions of deep structure and universal grammar.

†M. S. Seidenberg and L. A. Petitto state that "this kind of reaction," such as Patterson's, "trivializes the issues and does nothing to advance the debate."[499]

bly arose out of his train of thought starting from "universal doubt" to an assertion of his existence as "a thinking substance" in the famed *Cogito* statement.* Doubting his former beliefs allowed Descartes to reexamine and discard certain Scholastic ideas about physical substances as well as to replace an unquestioning reliance on authority (whether Aristotle or the medieval theologians) with a more empirical approach in conformity with reason.[500]

Nevertheless, despite his qualms about the Scholastics, Descartes strongly argued for the same metaphysical conclusions of the existence of man's imperishable soul and of a Perfect Being that both created and sustained it. These were "clear and distinct" ideas that lay beyond doubt for Descartes; yet he was indebted to lines of reasoning stemming from times past, i.e., the casual and ontological arguments for the existence of God, as seen throughout his *Discourse on Method* and *Meditations of First Philosophy;* for example:

> Perhaps, however, the being on which I am dependent is not the being I call God; perhaps I am produced by my parents, or by some other cause less perfect than God. But this cannot be true, for, as I have already said, it is obvious that there must be at least as much reality in the cause as in the effect, and since I am a thinking thing and have in me an idea of God, any cause that may be assigned to my nature must also be a thinking thing, and have in it the idea of all the perfections I attribute to God.[501]

Descartes argued that these clear and distinct ideas were innate, not attributable either to his sensory perception of the environment or to his own "invention" but rather to God: "And certainly I ought not to find it strange that God, in creating me, has placed this idea ["of a supremely perfect being"] in me to be, as it were, the mark of the workman imprinted on his work."[502]

Later rationalists accepted Descartes's methodological assumptions concerning the attainment of

---

Cogito, ergo sum ("I think, therefore I am."), Part IV, *Discourse on Method* in *Essential Works of Descartes,* trans. Lowell Bair (New York: Bantam, 1961), pp. 19–24.

knowledge, but without the "excess baggage" of theistic belief, however large or small the role of a divine Creator ultimately fit into Descartes's total system.[503]

For Descartes, the rational soul of man distinguished him from animals and machines (automata) in two respects: the ability to make ethical decisions in a diversity of situations and the ability to use language freely.[504] That Descartes could see the use of language as crucial evidence for the existence of other minds is indicative of his debt to the Scholastic grammarians, the Modistae, of the twelfth to the fourteenth centuries. The Modistae held that language as a reflection of thought was rule-governed because the universe as a whole was rule-governed (ordered).[505] The basis of the universal principles of grammar they proposed, therefore, was more metaphysical than psychological, that is, primarily based on a view of the reality of the external world.[506]

A century after Descartes wrote on the created soul and the use of language, there was a resurgence in the link made between a belief in a Creator and an appreciation for the regularity and complexity of language. The link emerged during a debate concerning the origin of language at the Berlin Academy of Sciences. Johann Suessmilch, a German church leader and innovative statistician, argued for the divine origin of language. Emphasizing the orderliness, complexity, and *universality* of language, Suessmilch ruled out both natural development and man's invention as possible causes. He argued that the design of language had to come from an intelligent source, the faculty of which man could not have had without having language in the first place. Having eliminated both chance and human invention, Suessmilch concluded in the rationalist style of Leibniz that there was only one alternative—the divine origin.[507]

Some years later, Johann Herder, a German philosopher and poet, wrote for the Berlin Academy his classic *Treatise on the Origin of Language* (in 1770). Herder attacked Suessmilch's position, while holding to similar views of the uniformity and complexity of language. Arguing for the *innateness of both human* *language and reason*, Herder maintained that language arose out of the "reflexiveness" of the human soul.[508] He took Suessmilch to task for some of his notions about language, which Herder thought invalidated Suessmilch's conclusion. One of Herder's main objections concerned Suessmilch's view that the sounds of all languages could be represented by some twenty letters of the alphabet, as an indication of a divinely planned economy.[509] Nevertheless, perhaps Herder's chief complaint against the divine origin theory was that it precluded any further investigation.*[510]

Mindful, then, of Herder's complaint, we now come to the point in this exposition where we consider the biblical account of man and language as another alternative—one that should no more be ruled out than should the holding of this view preclude fruitful investigation of language.

Another Look at an Old Source

In the biblical account, human beings are recorded to have been created specially by God to bear a certain "likeness" for a unique relationship that would necessarily involve both an order of mental organization beyond that of the animals and an ability to make use of articulate speech, as observed in the verbal exchanges recorded in Genesis 2–3. In chapter 11 of Genesis an account is offered of the beginnings of different languages: the diversification of the ancestral language accords with the *monogenetic* (single-source) view of language origin. Moreover, it is compatible with what we know of the essential underlying sameness of all human languages, as determined by a genetically specified set of universal principles.

Consequently, we may also put forth another alternative—one that is based not on a still unknown quantity of fortuitous occurrences but rather on other as-

---

*The reader is invited to consider the argument of Suessmilch, in Stam, *Inquiries*, chap. 5, notwithstanding the partially unfounded criticisms of Herder.

sumptions of the origin and maintenance of the universe, i.e., that an all-knowing and all-powerful Being designed humans with the ability to think and communicate those thoughts on a vastly different level from what animals would be capable of. The linguistic distinction between humans and animals (including the great apes) that the biblical account offers is thus part of a larger sphere of *qualitative differences* involving man's moral and personal fabric.

Obviously, these latter matters cannot be discussed within the confines of this book, but, if we may understate the matter, the account offered in Genesis is *not* incompatible with the facts of human language. As far as empirical studies now indicate, human language is the sole possession of a creature whose apparently insatiable sense of reason-for-being prompts him to continue wondering about himself in this universe.

If we may use one more metaphor, it is not unreasonable to bring down from the attic of discarded ideas of the Modistae and rationalists not just some dusted-off treasures, but also the full chest—the basis of such claims about man's nature and unique possession of language is the belief in the existence of a personal-infinite God. This account may therefore be offered as a coherent and viable alternative to the current speculation about man's evolutionary development.[511]

We conclude this study by quoting from psychologist-philosopher Mortimer Adler, who, in *The Difference of Man and the Difference it Makes,* states the importance and the far-reaching implications concerning a view of man's place in the world:

> The practical consequences of regarding man as differing only in degree from other animals all turn on the abrogation of the distinction we make between persons and things—a distinction that involves a difference in kind. The dignity of man is . . . a dignity that is not possessed by things. . . . The dignity of man as a person underlies the moral imperative that enjoins us never to use other human beings merely as means but always to respect them as ends to be served. . . . Hence, it would appear to make a great practical difference whether we can preserve the

distinction between men as persons and all else as things, or must abrogate it because men differ from all else only in degree.[512]

In awe of the almighty and loving Creator of the universe, the psalmist David says, "I praise you because I am fearfully and wonderfully made; your works are wonderful, I know that full well" (Psalm 139:14, NEW INTERNATIONAL VERSION).

Response

MARVIN K. MAYERS

Is there a difference between human speech and animal communication? To the average person it seems there is a difference. Moreover, to the believing Christian it seems there "ought" to be. Wilson and McKeon not only suggest there is, they also review for us experiments with chimps and apes that demonstrate clearly that there is a qualitative difference between the two. To various scientific hypotheses regarding the comparison of human speech with animal communication, the authors add one more—one that the evangelical world will gladly embrace and the so-called "secular" world should thoroughly consider. The authors call it a "creationist" view and speak of an all-knowing and powerful Being who designed humans with the ability to think and communicate their thoughts on a vastly different level from that of animals.

The chapters on language (3-4) and on language acquisition (6) alone are worth the cost of the entire book. The Christian will look a long way for such clear and incisive overviews of these respective fields. Those not endorsing a Christian view will look a long way for a viewpoint presented in a more scholarly or academic manner. Overall, the authors have presented a significant and truly scholarly work with the various arguments worked through carefully and thoroughly.

The most pertinent issue the authors raise in this work erupts on the final pages, where the question of human nature, a question latent throughout the text, arrives full-blown with all of its attendant implications. What the authors propose is no less than a radical reconsideration not only of the clearly qualitative differences between human and animal but also of alternative source material in the quest for a true understanding of human origins. Wilson and McKeon have suggested that we dust off formerly discarded ideas and examine the full biblical chest for answers to the language dilemma, human nature, and moral values.

The reader who is not committed to a Christian viewpoint, should try for a moment to suspend his disbelief and consider some of the implications of a Christian view of human nature. The quotation from Mortimer Adler at the end of the monograph makes it clear that what the authors have concluded—that man is *qualitatively* distinct from the animals—fits well with the biblical model. Such a view implies a dignity in humans not possessed by things. Now what sort of dignity is this?

From the biblical record we infer that man is endowed with a communicative ability—a personality—that reflects that of his Creator. The Genesis account indicates that man, not animal, holds a unique relation to God. The Creator's works are all wonderful, but this one, the human race, He made in His own image. Thus, the qualitative difference becomes clearer: humans may enjoy a personal relationship with the Creator God of the universe as an endowment from Him. The biblical account offers further the explanation that when this relationship is enjoyed and when the creature, so carefully designed, seeks peace with his Creator, then a relationship of incredible, nearly unspeakable, depth and satisfaction results.

Adler raises a further issue, that of moral imperative. Why is it that humans so often (even, we confess, Christians, who claim to serve the God of the Bible) *do not respect* the other humans who share this planet with them? Thus we have not only a moral imperative but also a moral dilemma. The Bible indicates that the free creatures violated their relationship with their Creator and that this violation has had far-reaching and devastating implications. The most basic of these implications is that the Creator God was forced to break off relations with the rebellious creature. While this response is no place to discuss the depths of the problem of good and evil, I must point out the two major premises usually missing from discussions of the subject. These are, first, that while there are problems attendant upon the Christian, or biblical, view of human nature and its implication that man is sinful, God has provided the answer to all those dilemmas by giving, of His own will, a Sacrifice that pleased His own nature and therefore allows people to know Him with the same satisfying depth enjoyed by the original humans. Second, competing world views such as pantheism or naturalism leave the one who considers their answers to questions of good and evil much less satisfied. In other words, answers to this problem from other world views entail problems of coherence, internal consistency, fragmentation, and psychological despair. The biblical world view, on the other hand, is coherent, consistent, full-orbed, and, most importantly, true and satisfying. There is a reason for this, as you will see toward the end of this response.

As the "curriculum world view" (inside the back cover of this book) claims, "the Creator has clothed Himself with humanity and become a part of the material realm in the person of Jesus of Nazareth, who died on a cross to remove our sins and was raised from the dead to authenticate His unique claims . . . that He has with finality dealt with man's moral flaw, that He desires to restore all people to a vital personal relationship with Him, and that any person may enter into and experience the happiness of fellowship with God by accepting through faith the death of Jesus as a substitute for his own." Thus has the Creator God dealt with man's flaw—with finality.

Finally, we have come full circle in considering the biblical view of human nature—from who man is, to who God is, to man's potential relationship with his Creator and the way that God has dealt with humanity's moral flaw. All that is left is the problem of communication. How do we know all this? Wilson and McKeon have demonstrated scientifically and accurately, I believe, the logic and reasoning behind human cognitive and communicative ability. If humans are endowed, therefore, with an innate and deep structure from which language naturally develops (barring injuries and deformities) in children, what does this imply about the issue of human origins and the Creator's communication with man?

I would like to develop the implications of this communication briefly at three levels, first philosophically; second, scientifically; and third, anthropologically.

First, a metaphor, or illustration, to point up a *philosophical* implication:[513]

> When an automobile manufacturer produces a new automobile, he also prepares an owner's manual. The manufacturer has designed and built the car; so he knows what type of fuel and lubricants are best for it. He knows the tire pressure that will give the best performance. He knows at what speeds it is best to shift gears. In short, the manufacturer knows better than anyone else how to operate the car. He has written the owner's manual, not to keep people from enjoying the car but to guide them in getting the most out of the car. God designed the universe and created man to live in it. Since God designed and created man, He knows best what man needs to function. God's "owner's manual" for the human race is the Bible. It was given not to keep us from enjoying life but to help us get the most out of life.

This metaphor illustrates the point of view called functional creationism. Each part of creation serves a purpose; creation is an integrated whole; and God's revelation of Himself, His own character, and His creation gives the human race the necessary understanding of *who* we are, the laws of the universe, and how we relate to those laws and our Creator. The purpose of the metaphor and explanation is to awaken some new *philosophical* perspective in the reader. God has created each person with not only an innate ability to communicate but, further, an innate knowledge of Himself (see Romans 1). In other words, the Creator has communicated with us, the human race, and has provided the means by which we may know Him, ourselves, others, and the rest of the Creation as He designed it. These truths allow us to live better in a world that too often refuses to acknowledge Him.

The second implication raised by man's communicative ability is *scientific*. You will note that in the text of *The Language Gap* the authors used normal scientific means to draw their conclusions: they gathered data, sifted through it, applied their expertise, and finally drew rational conclusions. Yet the conclusions they have drawn appear much "out of sync" with the conclusions of many intellectuals today. The reason should be obvious by now: the authors did not rule out God from the beginning of their research. Instead, they have recognized Him as Creator and as communicator with man *throughout* their research. This, then, is the *scientific* implication: man's innate communicative ability reflects the Creator God presented by the Bible. Yet the data of the Bible is all too often ignored and rejected as a viable resource for scientific assumptions. McKeon and Wilson have presented a rational case that the scientific community should consider. And they should weigh its implications.

The Bible itself points out that it is the nature of humans to suppress their knowledge of the Creator and to refuse to recognize Him (though this rejection is often the rejection of a *caricature* of God). See Romans 1:18–32. This perspective unfortunately dominates the scientific community today and thus militates against a full-orbed, coherent, and satisfying philosophy of science. The only way out of the dilemma that this scientific implication raises is for individuals within the intellectual community today to rethink their caricature of the Bible and the God it presents.

This leaves us with the third, *anthropological,* implication. A functional creationist viewpoint may underlie much of cultural anthropology. Anthropology views the various aspects of society as functional and society itself as an integrated whole, and functional creationism holds that God's rules for living are not capricious but functional. For example, God did not limit sexual activity to the family setting, between husband and wife, arbitrarily. His rules concerning human sexual activity are functional. After the Communist revolution in Russia in 1917, the new Communist leaders, thinking that rules pertaining to sex and fidelity were only part of a Christian ethic, encouraged premarital sex and enacted laws that made obtaining a divorce as easy as getting married. As they began to see their society disintegrate, they realized the rules for family maintenance were also sociologically sound. Today it is almost impossible to get a divorce in Russia, and promiscuity is officially discouraged. Thus the Communists discovered, through experience, the functional nature of God's proscription (though they don't recognize or acknowledge the source).

One of the biblical ideas that Wilson and McKeon suggest for consideration and that the anthropological community should wholeheartedly evaluate is that we live in a universe, not a multiverse. If we recognize that creation is unified and that each aspect of the creation is purposeful, or teleological, we may begin to be more considerate of the two previous implications—the philosophical one: that God has spoken to us, and the scientific one: that the biblical account should in no way limit our philosophy of science, but rather expand it. Thus our view of humanity and the social activities of man in a unified creation may lend a clearer and more helpful view of the diversities we find in world cultures.

Only anthropology, apart from the Bible, offers insight into the origin of man as distinct from animal. The form and expression of *Homo sapiens* is distinct from any simian or mammal or any other nonsapien primate that ever existed. It is this that marks man off from the rest of creation. Cro-Magnon man and those following (no matter when they lived) had all the mental, social, emotional, and spiritual potential spoken of in the early chapters of Genesis. There is evidence of art, metal working, and musical expression in the artifactual remains of the earliest forms of *Homo sapiens*.

Humans are set off from other creatures by their unique ability to symbolize through language and to develop social structures and organization. This leads to their ability to abstract spiritual insights. These abilities allowed them to know God as well as their fellowman. This human distinctiveness is referred to in the Bible as the "image of God." And this presupposition is what makes *The Language Gap* worthy of your consideration.

References

[1] Helen Keller, *The Story of My Life* (Garden City, N.Y.: Doubleday, 1954), pp. 34–35.

[2] Ibid., p. 35.

[3] Ibid., p. 36.

[4] Ibid., p. 257.

[5] Edward Sapir, "Language," in *Culture, Language and Personality: Selected Essays*, ed. David G. Mandelbaum (Berkeley: University of California Press, 1966), p. 15.

[6] E. Sue Rumbaugh and Duane M. Rumbaugh, "Symbolization, Language, and Chimpanzees: A Theoretical Reevaluation Based on Initial Language Acquisition Processes in Four Young Pan Troglodytes," *Brain and Language*, vol. 6, no. 3 (November 1978): 266.

[7] W. N. Kellogg and L. A. Kellogg, *The Ape and the Child: A Study of Environmental Influence on early behavior* (1933; reprint ed., New York: Hafner, 1967).

[8] Winthrop N. Kellogg, "Communication and Language in the Home-Raised Chimpanzee," *Science* 162: 426.

[9] See Cathy Hayes, *The Ape in Our House* (New York: Harper & Brothers, 1951).

[10] Kellogg, *The Ape and the Child*, p. 424.

[11] Philip Lieberman, *On the Origins of Language: An Introduction to the Evolution of Human Speech* (New York: MacMillan, 1975), chaps. 8–9.

[12] R. A. Gardner and B. T. Gardner, "Teaching Sign Language to a Chimpanzee," *Science*, 165 (1969): 664–72. Reprinted in R. L. Schiefelbusch and J. H. Hollis, eds. *Language Intervention from Ape to Child*, Language Intervention Series, vol. 3 (Baltimore: University Park Press, 1979), pp. 171–95.

[13] For a linguistic analysis, see E. Klima and U. Bellugi, *The Signs of Language* (Cambridge, Mass.: Harvard University Press, 1979).

[14] Gardner and Gardner, "Teaching Sign Language to a Chimpanzee," p. 176.

[15] Ibid., pp. 78–83.

[16] Ibid., p. 189.

[17] B. T. Gardner and R. A. Gardner, "Two-way Communication with an Infant Chimpanzee," in A. M. Schrier and F. Stollnitz, eds., *Behavior of Non-Human Primates*, vol. 4 (New York: Academic Press, 1971). Reported in A. Akmajian, R. A. Demers, and R. M. Harnish, *Linguistics: An Introduction to Language and Communication* (Cambridge, Mass.: M.I.T. Press, 1979), pp. 340–41. For more on children's stages, see chapter 5 of this work, part II; for more on the Gardners' analysis, see chapter 8.

[18] Gardner and Gardner, in Schiefelbusch and Hollis, *Language Intervention*, p. 191.

[19] Ibid., pp. 179–80.

[20] B. T. Gardner and R. A. Gardner, "Early Signs of Language in Child and Chimpanzee," in Schiefelbusch and Hollis, *Language Intervention*, pp. 199–204.

[21] This example with discussion is furnished by Herbert Terrace, *Nim: A Chimpanzee Who Learned Sign Language* (N.Y.: Washington Square, 1981), p. 15. See section V of this chapter. Also see an early discussion in Roger Brown, "The First Sentences of a Child and Chimpanzee" in T. A. Sebeok and J. Umiker-Sebeok, eds., *Speaking of Apes: A Critical Anthology of Two-Way Communication with Man* (New York: Plenum, 1980), pp. 85–101. Also discussed in chapter 8.

[22] "Update," *Newsweek*, 28 May 1979, p. 17. See also, Eugene Linden, "Endangered Chimps in the Lab," *New York Times Magazine* (December 19, 1982): 80.

[23] B. T. Gardner and R. A. Gardner, "Two Comparative Psychologists Look at Language Acquisition," in K. Nelson, ed., *Children's Language*, vol. 2 (New York: Gardner, 1980), pp. 336–37.

[24] A. J. Premack and D. Premack, "Teaching Language to an Ape," *Scientific American* (October 1972), 227(4): 92–99.

[25] Ibid., p. 92.

[26] Ibid., p. 95.

[27] Ibid.

[28] Ibid.

[29] Ibid., p. 96.

[30] The Gardners comment on the narrowness of this "forced-choice" test procedure in their "Evidence for Sentence Constituents in Early Utterances of Child and

Chimpanzee," *Journal of Experimental Psychology* 104 (1971): 244–67.

[31] A. J. Premack and D. Premack, "Teaching Language to an Ape," p. 95.

[32] Ibid., pp. 98–99.

[33] Duane Rumbaugh et al., "Reading and Sentence Completion by a Chimpanzee *(Pan),*" *Science* 182 (November 16, 1973): 731–33.

[34] Ibid., p. 733.

[35] Described by its designer, E. von Glaserfeld, in his article "The Yerkish Language and Its Automatic Parser," in D. M. Rumbaugh, ed., *Language Learning by a Chimpanzee: THE LANA PROJECT* (New York: Academic, 1977), pp. 91–130.

[36] Described in D. M. Rumbaugh and T. V. Gill: "Training Strategy and Tactics," pp. 157–62; idem, "Lana's Acquisition of Language Skills," pp. 165–92, in Rumbaugh, *Language Learning.*

[37] E. Sue Savage-Rumbaugh and Duane M. Rumbaugh, "Symbolization, Language, and Chimpanzees: A Theoretical Reevaluation Based on Initial Language Acquisition Processes in Four Young Pan Troglodytes," *Brain and Language,* vol. 6, no. 3 (November 1978): 265–300.

[38] Ibid., p. 275.

[39] Ibid., p. 273.

[40] Ibid., p. 278.

[41] Ibid., p. 283.

[42] Ibid., p. 286.

[43] Ibid., p. 298.

[44] Ibid.

[45] Ibid.

[46] Herbert S. Terrace, *Nim: A Chimpanzee Who Learned Sign Language* (New York: Washington Square, 1981. Originally published by New York: Knopf, 1979).

[47] Ibid., pp. 31–32.

[48] Noam Chomsky, *Language and Mind,* enlarged ed. (New York: Harcourt Brace Jovanovich, 1972).

[49] B. F. Skinner, *Verbal Behavior* (New York: Appleton Century Crofts, 1957).

[50] Terrace, *Nim,* p. 59.

[51]Ibid.
[52]Ibid., p. 260.
[53]Ibid., p. 270.
[54]Ibid., p. 317.
[55]Ibid., pp. 314, 319.
[56]Ibid., p. 316.
[57]Ibid.
[58]Francine G. Patterson, "The Gestures of a Gorilla: Language Acquisition in Another Pongid," *Brain and Language*, vol. 5, no. 1 (January 1978): 73 (see entire article, pp. 72–97).
[59]Francine G. Patterson, "Conversations With a Gorilla," *National Geographic* (October 1978), pp. 451, 453 (see entire article, pp. 438–465).
[60]Patterson, "Gestures of a Gorilla," p. 86.
[61]Ibid., p. 82.
[62]Ibid., p. 83.
[63]Ibid., pp. 83–84.
[64]Ibid., p. 88.
[65]Francine G. Patterson and Eugene Linden, *The Education of Koko*, (New York: Holt, Rinehart and Winston, 1981), p. 116.
[66]Robert Brown, *A First Language* (Cambridge, Mass.: Harvard University Press, 1973). See chapter 5.
[67]Patterson and Linden, *Education of Koko*, p. 118.
[68]Ibid., p. 117.
[69]Ibid.
[70]Patterson, "Gestures," p. 86. See also Patterson and Linden, *Education of Koko*, chaps. 16–17.
[70]Patterson, "Conversations," pp. 459–61.
[72]Ibid., p. 454; see also Patterson and Linden, *Education of Koko*, pp. 126–27.
[73]Ibid., p. 462.
[74]Patterson and Linden, *Education of Koko*, p. 109.
[75]Ibid., pp. 109–10.
[76]Patterson, "Gestures," p. 88.
[77]Patterson and Linden, *Education of Koko*, p. 140.
[78]Ibid.
[79]Ibid., p. 141.

References

[80] Patterson, "Gestures," p. 95.
[81] Ibid.
[82] Ibid.
[83] Eric H. Lenneberg, *Biological Foundations of Language* (New York: Wiley, 1976), p. 176.
[84] Edward Sapir, *Language: An Introduction to the Study of Speech* (New York: Harcourt, Brace and World, 1921), p. 22.
[85] Noam Chomsky, "Form and Meaning in Natural Languages," in *Language and Mind,* enlarged ed. (New York: Harcourt Brace Jovanovich, 1972).
[86] Noam Chomsky, *Aspects of the Theory of Syntax* (Cambridge, Mass.: M.I.T. Press, 1965), p. 8.
[87] Ibid., pp. 148–53.
[88] Albert C. Baugh and Thomas Cable, *A History of the English Language,* 3rd ed. (Englewood Cliffs, N.J.: Prentice-Hall, 1978), chap. 9.
[89] Diane D. Bornstein, *Readings in the Theory of Grammar* (Cambridge, Mass.: Winthrop, 1976), chap. 1. Also see Julia S. Falk, *Linguistics and Language: A Survey of Basic Concepts and Implications,* 2nd ed. (New York: Wiley, 1978), chaps. 1–2.
[90] William Labov, "The Logic of Nonstandard English" (corrected version of original article in *Georgetown 20th Annual Round Table Report,* James E. Alatis, ed., 1969) in Richard W. Bailey and Jay L. Robinson, eds., *Varieties of Present-Day English* (New York: Macmillan, 1973).
[91] William Labov, *The Study of Nonstandard English* (Champaign, Ill.: N.C.T. E., 1970), pp. 40–41.
[92] Walt Wolfram, "Black-White Speech Differences Revisited," *Black-White Speech Relationships,* ed. Clarke and Wolfram (Washington, D.C.: Center for Applied Linguistics, 1971), p. 156.
[93] Labov, "The Logic of Nonstandard English," in Bailey and Robinson, *Varieties,* p. 351.
[94] Walt Wolfram and Donna Christian, *Appalachian Speech* (Washington, D.C.: Center for Applied Linguistics, 1976).
[95] Labov, *The Study of Nonstandard English,* p. 17.
[96] Noam Chomsky, *Language and Responsibility* (New York: Pantheon, 1979), pp. 165–68.

[97] Ray S. Jackendoff, *Semantic Interpretation in Generative Grammar* (Cambridge, Mass.: M.I.T. Press, 1972), chap. 2.

[98] Noam Chomsky, *Reflections on Language* (New York: Pantheon, 1975), pp. 141–42.

[99] Chomsky, *Language and Mind*, pp. 11–12.

[100] Ibid., p. 100.

[101] *Aspects of the Theory of Syntax*, pp. 4–5.

[102] Noam Chomsky, *Rules and Representations* (New York: Columbia University Press, 1980), pp. 223–26.

[103] Helen S. Cairns and Charles E. Cairns, *Psycholinguistics: A Cognitive View of Language* (New York: Holt, Rinehart and Winston, 1976), p. 38.

[104] Noam Chomsky and Morris Halle, *The Sound Pattern of English* (New York: Harper & Row, 1968), pp. 24–25. Also see Philip Lieberman, *Intonation, Perception, and Language*, Research Monograph No. 38 (Cambridge, Mass.: M.I.T. Press, 1967), pp. 162–67.

[105] Chomsky and Halle, *The Sound Pattern of English*, pp. 293–95. Also see Philip Lieberman, *Speech Physiology and Acoustic Phonetics: An Introduction* (New York: Macmillan, 1977), pp. 117–20.

[106] Eric H. Lenneberg, *Biological Foundations of Language* (New York: Wiley, 1976), pp. 91–92.

[107] Noam Chomsky, *Language and Responsibility* (New York: Pantheon, 1979), chapter 6.

[108] Chomsky, *Rules and Representations*. pp. 224–25.

[109] Chomsky, *Aspects of the Theory of Syntax*, p. 4.

[110] Ibid. The model is the general theme of *Aspects*.

[111] Chomsky, *Reflections on Language*.

[112] Chomsky, *Rules and Representations*, pp. 196–97.

[113] Adrian Akmajian and Fred W. Henry, *An Introduction to the Principles of Transformational Syntax* (Cambridge, Mass.: M.I.T. Press, 1975).

[114] Peter W. Culicover, *Syntax* (New York: Academic Press, 1976). (Now in 2nd ed., 1982.)

[115] C. L. Baker, *Introduction to Generative-Transformational Syntax* (Englewood Cliffs, N.J.: Prentice-Hall, 1978).

[116] J. A. Fodor, T. Bever, and M. Garrett, *The Psychology of Language: An Introduction to Psycholinguistics and*

[117] Noam Chomsky and Morris Halle, *The Sound Pattern of English* (New York: Harper & Row, 1968), p. 297.

[118] Ronald W. Langacker, *Language and Its Structure: Some Fundamental Linguistic Concepts,* 2nd ed. (New York: Harcourt Brace Jovanovich, 1973).

[119] A. Akmajian, R. A. Demers, and R. M. Harnish, *Linguistics: An Introduction to Language and Communication* (Cambridge, Mass.: M.I.T. Press, 1979).

[120] Victoria Fromkin and Robert Rodman, *An Introduction to Language,* 2nd ed., (New York: Holt, Rinehart and Winston, 1978).

[121] C. Sloat, S. H. Taylor, and J. E. Hoard, *Introduction to Phonology* (Englewood Cliffs, N.J.: Prentice-Hall, 1978).

[122] Chomsky and Halle, *Sound Pattern of English.*

[123] Michael K. Brame, *Conjectures and Refutations* (New York: North Holland, 1976).

[124] Ray Jackendoff, "Grammar as Evidence for Conceptual Structure," in M. Halle, J. Bresnan, and G. Miller, eds., *Linguistic Theory and Psychological Reality* (Cambridge, Mass.: M.I.T. Press, 1978), pp. 201-28. Also see: J. D. Fodor, J. A. Fodor, and M. F. Garrett, "The Psychological Unreality of Semantic Representations," *Linguistic Inquiry,* no. 6 (1975), pp. 515-31; and N. Chomsky, *Language and Responsibility,* chapter 6.

[325] Noam Chomsky, *Essays on Form and Interpretation* (New York: North Holland, 1976).

[126] Akmajian, Demers, and Harnish, *Linguistics,* p. 242.

[127] Ibid., chap. 7.

[128] J. A. Fodor and J. J. Katz, "The Structure of a Semantic Theory," *Language,* 39 (1963): 170-210. Reprinted in J. A. Fodor and J. J. Katz, *The Structure of Language* (Englewood Cliffs, N.J.: Prentice-Hall, 1964). See also Jerrold J. Katz, *Semantic Theory* (New York: Harper & Row, 1972).

[129] Ruth M. Kempson, *Semantic Theory* (Cambridge: Cambridge University Press, 1977), pp. 18-20.

[130] John Lyons, *Semantics,* vols. 1 and 2 (Cambridge: Cambridge University Press, 1977), pp. 317-35.

[131] Paul Kiparsky and Carol Kiparsky, "Fact," in D. Steinburg and L. Jakobovits, eds., *Semantics: An Interdisciplinary Reader in Philosophy, Linguistics, and Psychology* (New York: Cambridge University Press, 1971), pp. 345-69.

[132] H. Paul Grice, "Logic and Conversation," in P. Cole and J. Morgan, eds., *Syntax and Semantics*, vol. 3 (New York: Academic, 1975), pp. 45-47. Also, see Kempson, *Semantic Theory*, pp. 68-73.

[133] Ruth Kempson, *Presupposition and the Delimitation of Semantics* (Cambridge: Cambridge University Press, 1975), chap. 8.

[134] Lyons, *Semantics*, vols. 1 and 2; Kempson, *Semantic Theory;* F. R. Palmer, *Semantics: A New Outline* (Cambridge: Cambridge University Press, 1976); Janet D. Fodor, *Semantics: Theories of Meaning in Generative Grammar* (New York: Crowell, 1977).

[135] Lenneberg, *Biological Foundations of Language*, p. 176.

[136] Norman Geschwind, "Language and the Brain," *Scientific American* (April 1972): 76-83.

[137] Ibid., p. 76.

[138] Lenneberg, *Biological Foundations of Language*, pp. 151-52.

[139] Geschwind, "Language and the Brain," p. 76.

[140] Ibid., p. 78.

[141] Ibid., p. 79.

[142] John C. Eccles, *The Understanding of the Brain* (New York: McGraw-Hill, 1973), pp. 201-4.

[143] Ibid., p. 204.

[144] Michale S. Gazzaniga, *The Bisected Brain* (New York: Appleton-Century-Crofts, 1970).

[145] Doreen Kimura, "The Assymetry of the Human Brain," *Scientific American* (March 1973): 70.

[146] Ibid.

[147] Lenneberg, *Biological Foundations of Language*, p. 178.

[148] Ibid., pp. 175-76.

[149] Ibid., pp. 142-50.

[150] Jill deVilliers and Peter deVilliers, *Language Acquisition* (Cambridge, Mass.: Harvard University Press, 1978), pp. 210-11, citing Stephen Krashen, "Lateralization,

Language Learning, and the Critical Period: Some New Evidence," *Language Learning* 23 (June 1973): 63–74.

[151] For example, see John Schumann and Nancy Stenson, eds., *New Frontiers in Second Language Learning* (Rowley, Mass.: Newbury, 1974); Heidi Dulay, Marina Burt, and Stephen Krashen, *Language Two* (New York: Oxford University Press, 1982).

[152] Lenneberg, *Biological Foundations of Language*, p. 151.

[153] L. S. Basser, "Hemiplegia of Early Onset and the Faculty of Speech with Special Reference to the Effects of Hemispherectomy," *Brain* 85 (1962): 427–60.

[154] Lenneberg, *Biological Foundations of Language*, p. 151.

[155] Mildred Freburg-Berry, *Language Disorders of Children: The Bases and Diagnosis* (Englewood Cliffs, N.J.: Prentice-Hall, 1969), p. 42.

[156] Maureen Dennis and Harry Whitaker, "Language Acquisition following Hemidecortisation: Linguistic Superiority of the Left over the Right Hemisphere," *Brain and Language* 3 (1976): 404–33; Maureen Dennis, "Language Acquisition in a Single Hemisphere: Semantic Organization," in *Biological Studies of Mental Processes*, ed. David Caplin (Cambridge, Mass.: M.I.T. Press, 1980), pp. 159–85; and Noam Chomsky, *Rules and Representations*, pp. 56–57.

[157] Lenneberg, *Biological Foundations of Language*, pp. 127–35. See the tables on pp. 128–34.

[158] Ibid., pp. 128–30.

[159] Ibid., pp. 126–31.

[160] Akmajian, Demers, and Harnish, *Linguistics*.

[161] Fromkin and Rodman, *Introduction to Language*.

[162] See Melissa Bowerman, *Early Syntactic Development: A Cross Linguistic Study With Special Reference to Finnish* (Cambridge: Cambridge University Press, 1973).

[163] Cairns and Cairns, *Psycholinguistics*, p. 193.

[164] Carol Chomsky, *The Acquisition of Syntax in Children from 5 to 10* (Cambridge, Mass.: M.I.T. Press, 1969).

[165] Roger Brown, *A First Language: The Early Stages* (Cambridge, Mass.: Harvard University Press, 1973).

[166] Ibid., pp. 53–54.

[167] Ibid., p. 272.

[168] Cairns and Cairns, *Psycholinguistics*, pp. 193–97. Also see: Fromkin and Rodman, *An Introduction to Language*, pp. 244–50; Philip S. Dale, *Language Development: Structure and Function*, 2nd ed. (New York: Holt, Rinehart and Winston, 1976).

[169] As expressed, for example, in Fromkin and Rodman, *Introduction to Language*, p. 245.

[170] Eve Clark, "What's in a Word? On the Child's Acquisition of Semantics in His First Language," in *Cognitive Development and the Acquisition of Language*, ed. Timothy E. Moore (New York: Academic, 1973), pp. 65–110.

[171] Susan Carey, "The Child as Word Learner," in M. Halle, J. Brennan, and G. Miller, eds., *Linguistic Theory and Psychological Reality* (Cambridge, Mass.: M.I.T. Press, 1978), pp. 264–93.

[172] See, for example, Roger Brown, *Words and Things* (New York: Free Press, 1958); T. E. Moore, ed., *Cognitive Development and the Acquisition of Language;* N. Chomsky, *Problems of Knowledge and Freedom* (New York: Pantheon, 1971), chap. 1; idem, *Reflections on Language*, pp. 44–52.

[173] Clark, "What's in a Word?" p. 76.

[174] Manfred Bierwisch, "On Classifying Semantic Features," *Progress in Linguistics*, ed. M. Bierwisch and K. Keidolph (The Hague: Mouton, 1970), pp. 27–50.

[175] David McNeill, *The Acquisition of Language* (New York: Harper and Row, 1970), pp. 20–25.

[176] DeVilliers and deVilliers, *Language Acquisition*, pp. 48–52.

[177] Roman Jakobson and Morris Halle, *Fundamentals of Language* (The Hague: Mouton, 1956), chap. 4.

[178] Ibid., p. 37.

[179] As noted in deVilliers and deVilliers, *Language Acquisition*, pp. 38–39. Also see: Phillip S. Dale, *Language Development, Structure and Function*, 2nd ed. (New York: Holt, Rinehart & Winston, 1976), pp. 212–13.

[180] N. V. Smith, *The Acquisition of Phonology: A Case Study* (Cambridge: Cambridge University Press, 1973); deVilliers and deVilliers, *Language Acquisition*, p. 41.

[181] Brown, *First Language*.

[182] Martin Braine, "The Ontogeny of English Phrase Structure: The First Phase," *Language* 39 (1963): 1–13.

References

[183] David McNeill, "Developmental Psycholinguistics," in F. Smith and G. Miller, eds., *The Genesis of Language: A Psycholinguistic Approach* (Cambridge, Mass.: M.I.T. Press, 1966), p. 22.

[184] Lois Bloom, *Language Development: Form and Function in Emerging Grammar* (Cambridge, Mass.: M.I.T. Press, 1970); Melissa Bowerman, *Early Syntactic Development—A Cross Linguistic Study with Special Reference to Finnish* (Cambridge: Cambridge University Press, 1973).

[185] Bowerman, *Early Syntactic Development*, p. 218.

[186] Brown, *First Language*, p. 173.

[187] Example comes from Braine, "The Ontogeny of English Phrase Structure," cited in Dale, *Language Development*, p. 22.

[188] Ursulla Bellugi, reported in deVilliers and deVilliers, *Language Acquisition*, pp. 102–5.

[189] Ibid., p. 103.

[190] Brown, *First Language*, p. 271.

[191] Discussed in deVilliers and deVilliers, *Language Acquisition*, pp. 99–100; Dale, *Language Development*, chap. 5.

[192] Fodor, Bever, and Garrett, *Psychology of Language*, pp. 319–28.

[193] Chomsky, *Aspects*, p. 9.

[194] Thomas Bever, "The Cognitive Basis for Linguistic Structures," in *Cognition and the Development of Language*, ed. J. R. Hayes (New York: Wiley, 1970); Dan Slobin, "Cognitive Prerequisites for the Development of Grammar," in *Studies of Child Language Development*, ed. C. A. Ferguson and D. I. Slobin (New York: Holt, Rinehart, and Winston, 1973), pp. 183–86; Fodor, Bever, and Garrett, *Psychology of Language*, chaps. 5–7; Cairns and Cairns, *Psycholinguistics*, chaps. 6 and 7.

[195] E.g., Joan Bresnan, "A Realistic Transformational Grammar," in Halle, Brennan, and Miller, *Linguistic Theory and Psychological Reality*, pp. 1–59.

[196] Dale, *Language Development*, pp. 34–36.

[197] Brown, *First Language*.

[198] Ibid., p. 272.

[199] Ibid., p. 271.

[200] Ibid., p. 377.

201. See deVilliers and deVilliers, *Language Acquisition*, p. 377.
202. Brown, *First Language*, p. 377.
203. Ibid., p. 273.
204. Carol Chomsky, *The Acquisition of Syntax in Children from 5 to 10* (Cambridge, Mass.: M.I.T. Press, 1969).
205. See William Labov, *The Study of Nonstandard English* (Champaign, Ill.: N.C.T.E., 1970), as opposed to the view of Noam Chomsky in *Language and Responsibility*, chap. 2.
206. deVilliers and deVilliers, *Language Acquisition*, pp. 199–203.
207. Ibid.
208. Ibid., p. 201.
209. Ibid., pp. 199–203.
210. Dale, *Language Development*, pp. 116–17.
211. Ibid., p. 138.
212. Brown, *First Language*, pp. 362–68.
213. Ibid., pp. 367–68.
214. Martin D. S. Braine, "The Ontogeny of English Phrase Structure," pp. 1–13.
215. T. G. Bever, J. A. Fodor, and W. Weksel, "On the Acquisition of Syntax: A Critique of 'Contextual Generalization,'" in L. A. Jakobovits and M. S. Miron, eds., *Readings in the Psychology of Language* (Englewood Cliffs, N.J.: Prentice-Hall, 1967); responded to by M. D. S. Braine, "On the Basis of Phrase Structure: A Reply to Bever, Fodor, and Weksel," in the same volume.
216. Noam Chomsky, *Cartesian Linguistics* (New York: Harper & Row, 1966), p. 12.
217. deVilliers and deVilliers, *Language Acquisition*, p. 207.
218. Ibid., p. 206. Also see Fromkin and Rodman, *Introduction to Language*, pp. 250–51.
219. N. Chomsky, "A Review of B. F. Skinner's *Verbal Behavior*," *Language* 35 (1959): 26–58. Reprinted in Fodor and Katz, *Structure of Language*, pp. 547–78.
220. McNeill, "Developmental Psycholinguistics," p. 29.
221. Jerry Fodor, "How to Learn to Talk: Some Simple Ways," in Smith and Miller, *Genesis of Language*, pp. 105–22.
222. Chomsky, *Cartesian Linguistics*, pp. 4–5.

[223] Ibid., p. 65.
[224] Chomsky, *Aspects of the Theory of Syntax*, pp. 47–54.
[225] Ibid., pp. 27–30.
[226] Chomsky, *Rules and Representations*, p. 232.
[227] Ibid., pp. 5, 39.
[228] Massimo Piatelli-Palmarini, ed., *Language and Learning: The Debate between Jean Piaget and Noam Chomsky* (Cambridge, Mass.: Harvard University Press, 1980), pp. 39–40.
[229] Chomsky, *Reflections on Language*, p. 118.
[230] Ibid., p. 4.
[231] Jean Piaget, "The Psychogenesis of Knowledge and Its Epistemological Significance," in Piatelli-Palmarini, *Language and Learning*, p. 23.
[232] Jean Piaget, "Introductory Remarks: About the Fixed Nucleus and Its Innateness," in Piatelli-Palmarini, *Language and Learning*, p. 58.
[233] H. Sinclair-deZwart, "Language Acquisition and Cognitive Development," in Timothy E. Moore, ed., *Cognitive Development and the Acquisition of Language* (New York: Academic Press, 1973), (pp. 9–25), p. 13.
[234] Donald M. Moorehead and Ann Moorehead, "From Signal to Sign: A Piagetian View of Thought and Language During the First Two Years," in *Language Perspectives—Acquisition, Retardation and Intervention* (Baltimore: University Park Press, 1974), (pp. 153–90), p. 155.
[235] Ibid., p. 154.
[236] Ibid., p. 155.
[237] Brown, *First Language*, pp. 362–68.
[238] David McNeill, *The Conceptual Basis of Language* (Hillsdale, N.J.: Lawrence Erlbaum Associates, 1979).
[239] Moorehead and Moorehead, "From Signal to Sign," pp. 183 and 170.
[240] Ibid., pp. 158–77; these pages summarize the six stages within the sensorimotor period.
[241] Jean Piaget, *The Language and Thought of the Child* (Cleveland: World, Meridian, 1955).
[242] Jerry Fodor, "Fixation of Belief and Concept Acquisition," in Piatelli-Palmarini, *Language and Learning*, pp. 143–49; idem, *The Language of Thought* (Cambridge, Mass.: Harvard University Press, 1979), pp. 87–97.

243. Fodor, "Fixation of Belief," p. 144.
244. Ibid., p. 145.
245. Ibid.
246. Ibid., p. 146.
247. Ibid.
248. Ibid., p. 148.
249. Chomsky, *Language and Responsibility*, pp. 84–85; idem, *Rules and Representations*, pp. 207–8.
250. Chomsky, *Rules and Representations*, p. 208.
251. Ibid., p. 236.
252. Jean Piaget, discussion of Fodor's paper, "Fixation of Belief," in Piatelli-Palmarini, *Language and Learning*, p. 150.
253. Ibid.
254. Ibid.; idem, "The Psychogenesis of Knowledge and Its Epistemological Significance," in Piatelli-Palmarini, *Language and Learning*, p. 26.
255. Jerry Fodor, discussion of "Fixation of Belief," p. 151.
256. Ibid., p. 152.
257. Ibid., p. 151.
258. Edward Sapir, "The Psychological Reality of the Phoneme," in *Selected Writings of Edward Sapir in Language, Culture, and Personality*, David G. Mandelbaum, ed. (Berkeley: University of California Press, 1966), pp. 46–60.
259. Edward Sapir, "Language," in David G. Mandelbaum, ed., *Culture, Language and Personality: Selected Essays* (Berkeley: University of California Press, 1949), pp. 1–44.
260. Ibid., pp. 15–16.
261. Ibid., p. 19.
262. Edward Sapir, "Anthropology and Psychiatry," in Mandelbaum, *Culture*, p. 151.
263. Ibid., pp. 151, 153.
264. Ibid., pp. 146–47.
265. Ibid.
266. Chomsky, *Language and Mind*, pp. 1–2, and elsewhere.
267. Chomsky, *Rules and Representations*, p. 208.
268. Edward Sapir, *Language* (New York: Harcourt, Brace & World, 1921, 1949), pp. 218–19.

[269] Sapir, "Language," p. 8.

[270] Benjamin Lee Whorf, *Language, Thought, and Reality: Selected Writings of Benjamin Lee Whorf*, ed. John B. Carroll (Cambridge, Mass.: M.I.T. Press, 1956), (pp. 246–70), p. 252.

[271] Joseph H. Greenberg, ed., *Universals of Language*, 2nd ed. (Cambridge, Mass.: M.I.T. Press, 1966).

[272] For strong criticism of Whorf, see R. W. Langacker, *Language and Its Structure*, pp. 36–41. For more moderate criticism, see John Lyons, *Language and Linguistics* (Cambridge, England: Cambridge University Press, 1981), pp. 304–17.

[273] Brent Berlin and Paul Kay, *Basic Color Terms: Their Universality and Evolution* (Berkeley: University of California Press, 1969).

[274] John Lyons, *Language and Linguistics*, pp. 314–17; also, F. C. Southworth and C. J. Daswani, *Foundations of Linguistics* (New York: Free Press, 1974), pp. 195–97.

[275] Dan Sperber, "Remarks on the Lack of Positive Contributions from Anthropologists to the Problem of Innateness" in Piatelli-Palmarini, *Language and Learning*, pp. 245–49.

[276] Ibid., pp. 248–49.

[277] Further introductory reading on the subjects of how linguistic competence is acquired and the nature of language universals, as well as the debate between rationalism and empiricism, is available; for example, Ronald W. Langacker, *Language and Its Structure*, chap. 9. Also, Julia S. Falk, *Linguistics and Language: A Survey of Basic Concepts and Implications*, 2nd ed. (New York: Wiley, 1978), chap. 22; McNeill, *Acquisition of Language*, chap. 5; Dale, *Language Development*, chap. 4; Noam Chomsky, *Language and Mind*, enlarged ed. (New York: Harcourt Brace Jovanovich, 1972), chap. 1.

[278] Noam Chomsky, *Language and Mind*, p. 69.

[279] Akmajian, Demers and Harnish, *Linguistics: An Introduction*, p. 63.

[280] Ibid., pp. 20–21, taken from W. Thorpe, *Bird Song* (Cambridge: Cambridge University Press, 1961).

[281] Ibid., p. 61.

[282] Ibid., p. 17, taken from K. von Frisch, *The Dance Language and Orientation of Bees*, trans. C. E. Chadwick (Cambridge, Mass.: Harvard University Press, 1967).

[283] Chomsky, *Language and Mind,* pp. 69–70.
[284] Langacker, *Language and Its Structure,* p. 19.
[285] Ibid., p. 20.
[286] Chomsky, *Language and Mind,* p. 70.
[287] Akmajian, Demers, and Harnish, *Linguistics: An Introduction,* chap. 4.
[288] Ibid., p. 63.
[289] Langacker, *Language and Its Structure,* 2nd ed., p. 19.
[290] McNeill, *Acquisition of Language,* p. 40; from P. Marler and W. V. Hamilton, *Mechanisms of Animal Behavior* (New York: Wiley, 1966); T. E. Rowell and R. A. Hinde, "Vocal Communication by the Rhesus Monkey (*Macaca Mulatta*)," *Proceedings of the Zoological Society of London* 138 (1962): 279–94.
[291] Lieberman, *On the Origins of Language,* chaps. 8–9.
[292] Akmajian, Demers, and Harnish, *Linguistics: An Introduction,* p. 33. Also see Thelma Rowell, *Social Behavior of Monkeys* (Baltimore: Penguin, 1972), p. 84.
[293] Peter Marler, "Vocal Ethology of Primates; Implications for Psycho-physics and Psychophysiology," in D. J. Chivers and J. Herbert, eds., *Recent Advances in Primatology* (New York: Academic, 1978), p. 795–801 (see especially p. 797).
[294] Akmajian, Demers, and Harnish, *Linguistics: An Introduction,* p. 33.
[295] Chomsky, *Language and Mind,* pp. 66–71.
[296] Charles Hockett, "The Origin of Speech," *Scientific American* 203 (1960): 88–96.
[297] William Thorpe, *Animal Nature and Human Nature* (Garden City, N.Y.: Doubleday, 1974), pp. 66–76.
[298] Akmajian, Demers, and Harnish, *Linguistics: An Introduction,* pp. 53–57.
[299] Chomsky, *Rules and Representations,* p. 241 and passim.
[300] Chomsky, *Language and Mind,* p. 70.
[301] Langacker, *Language and Its Structure,* pp. 245–46.
[302] R. A. Gardner and B. T. Gardner, "Comparative Psychology and Language Acquisition," in Sebeok and Umiker-Sebeok, *Speaking of Apes,* p. 289.
[303] Ibid., p. 293.

304Gardner and Gardner, "Two Way Communication with an Infant Chimpanzee," p. 118.

305Premack and Premack, "Teaching Language," p. 92.

306Ibid., p. 99.

307Ibid., pp. 98-99.

308D. Rumbaugh and T. V. Gill, "Lana's Acquisition of Language Skills," in Rumbaugh, ed., *Language Learning by a Chimpanzee* (New York: Academic, 1977), p. 190.

309Ibid., p. 191.

310Francine Patterson, "Gestures of a Gorilla," in *Brain and Language* 5 (1978): 95.

311F. Patterson and Eugene Linden, *The Education of a Gorilla*, p. 94.

312Ibid., p. 210.

313Patterson, "Gestures of a Gorilla," p. 95.

314Chomsky, *Language and Mind*, pp. 66-67. (The original three lectures, chaps. 1-3, were given at the University of California-Berkeley in 1967.)

315Ibid., p. 67.

316Chomsky, *Rules and Representations*, p. 239.

317Noam Chomsky, "Human Language and Other Semiotic Systems," in Sebeok and Umiker-Sebeok, eds., *Speaking of Apes*, p. 439.

318Ibid., pp. 434-35.

319Roger Brown, "The First Sentences of Child and Chimpanzee," in Sebeok and Umiker-Sebeok, eds., *Speaking of Apes*, pp. 85-101, originally published in his *Psycholinguistics: Selected Papers by Roger Brown* (New York: Free Press, 1970).

320H. S. Terrace et al., "Can an Ape Create a Sentence?" *Science* 206 (November 23, 1969), pp. 895, 899.

321Gardner and Gardner, "Two Comparative Psychologists Look at Language Acquisition," p. 354.

322Here is a series of references addressing the problem of rich interpretation:

L. A. Petitto and M. S. Seidenberg, "On the Evidence for Linguistic Abilities in Signing Apes," *Brain and Language* 8 (1979): 162-83.

M. S. Seidenberg and L. A. Petitto, "Ape Signing:

Problems of Method and Interpretation," in *The Clever Hans Phenomenon*, pp. 115–29.

Terrace, *Nim*, chap. 2.

Terrace, "Can an Ape Create a Sentence?"

Terrace, "On the Grammatical Capacity of Apes," in K. E. Nelson, ed. *Children's Language*, vol. 2 (New York: Gardner, 1980): pp. 371–495.

Terrace, "A Report to an Academy, 1980," in T. A. Sebeok and R. Rosenthal, eds., *The Clever Hans Phenomenon: Communication with Horses, Whales, Apes and People*, vol. 364 (New York Academy of Sciences, 1981): pp. 94–114.

Terrace, "Is Problem-Solving Language?" in T. A. Sebeok and Jean Umiker-Sebeok, eds., *Speaking of Apes: A Critical Anthology of Two-Way Communication with Man* (New York: Plenum, 1980), pp. 385–405.

Terrace et al., "Reply to Dalbir Bindra and Francine Patterson," *Science* 211 (Jan. 2, 1981): 86–87.

[323]Terrace, "Can an Ape Create a Sentence?" idem, p. 896; "Grammatical Capacity," pp. 422–26.

[324]Terrace, "Reply," p. 88.

[325]T. A. Sebeok and J. Umiker-Sebeok, 1980, pp. 14–15; "Questioning Apes," in Sebeok and Umiker-Sebeok, *Speaking of Apes*, pp. 1–59.

[326]Terrace, "Can an Ape Create a Sentence?" p. 899.

[327]Seidenberg and Petitto, "Ape Signing," p. 124. Lack of data is a matter we will discuss separately below.

[328]Ibid.

[329]E. Sue Savage-Rumbaugh, D. M. Rumbaugh, and S. Boysen, "Do Apes Use Language?" *American Scientist* (Jan.-Feb. 1980): 52.

[330]Seidenberg and Petitto, "Ape Signing," p. 116.

[331]Terrace, "Can an Ape Create a Sentence?" p. 896; idem, "Grammatical Capacity," pp. 424–25.

[332]Terrace, "Grammatical Capacity," pp. 425–26.

[333]Ibid., p. 425.

[334]Terrace, "Can an Ape Create a Sentence?" "Grammatical Capacity," "Reply," "Report," "Problem-Solving"; Seidenberg and Petitto, "Ape Signing"; Petitto and Seidenberg, "On the Evidence."

[335]Savage-Rumbaugh and Rumbaugh, "Symbolization, Language, and Chimpanzees: A Theoretical Reevalua-

tion Based on Initial Language Acquisition Processes in Four Young *Pan Troglodytes,*" *Brain and Language* 6 (1978): 278; Savage-Rumbaugh, Rumbaugh, and Boysen, "Do Apes Use Language?" pp. 51–52.

[336] Seidenberg and Petitto, "Ape Signing," pp. 120–21.

[337] Cited in Koko's case, Petitto and Seidenberg, "On the Evidence," p. 165.

[338] Seidenberg and Petitto, "Ape Signing," p. 120.

[339] Chomsky, "Human Language," p. 437.

[340] Ibid.

[341] Gardner and Gardner, "Comparative Psychology," pp. 313–18.

[342] Ibid., p. 317.

[343] Ibid., pp. 317–18.

[344] Chomsky, "Human Language," p. 437.

[345] Gardner and Gardner, "Comparative Psychology," p. 318; in the section on data-recording problems, pages 134–39, we further scrutinize the Gardners' claim of test proof methods for their strong conclusions, specifically their "double blind" test and their *wh*-Question test.

[346] Terrace, "Can an Ape Create a Sentence?" p. 899.

[347] Savage-Rumbaugh and Rumbaugh, "Symbolization"; Savage-Rumbaugh, Rumbaugh, and Boysen, "Do Apes Use Language?"

[348] Savage-Rumbaugh and Rumbaugh, "Symbolization," p. 278. See chap. 2 above.

[349] Ibid.

[350] Ibid.

[351] Ibid.

[352] Savage-Rumbaugh, Rumbaugh, and Boysen, "Do Apes Use Language?" p. 51.

[353] Ibid., p. 60.

[354] Petitto and Seidenberg, "On the Evidence," pp. 168–69.

[355] Ibid., p. 169.

[356] Sebeok and Umiker-Sebeok, "Questioning Apes," p. 15.

[357] Ibid.; compare Patterson's examples from chaps. 16 and 17 in Patterson and Linden, *Education of Koko,* mentioned in chap. 2.

358Terrace, "Can an Ape Create a Sentence?" pp. 895-96; Sebeok and Umiker-Sebeok, "Questioning Apes," p. 15.

359Seidenberg and Petitto, "Ape Signing," p. 120; Petitto and Seidenberg, "On the Evidence," pp. 164-65; Savage-Rumbaugh and Rumbaugh, "Symbolization," p. 278.

360Terrace, "Can an Ape Create a Sentence?" p. 900.

361Terrace, "Can an Ape Create a Sentence?" "Grammatical Capacity," "Reply," "Report"; Seidenberg and Petitto, "Ape Signing"; Petitto and Seidenberg, "On the Evidence."

362Seidenberg and Petitto, "Ape Signing," p. 120.

363Petitto and Seidenberg, "On the Evidence," p. 117.

364Terrace, *Nim*, pp. 321, 331-32.

365Terrace, "Grammatical Capacity," p. 425.

366Ibid.

367See chap. 5, pp. 80-84.

368Terrace, "Can an Ape Create a Sentence?" pp. 896-97, referring to data from Bloom et al., 1976. L. Bloom, L. Rocissano, and L. Hood, "Adult-Child Discourse: Developmental Interaction between Information Processing and Linguistic Knowledge," *Cognitive Psychology* 8 (1976): 521-22.

369Terrace, "Can an Ape Create a Sentence?" pp. 897-98.

370Ibid., pp. 897-98.

371Umiker-Sebeok and Sebeok, "Questioning Apes"; Sebeok and Rosenthal, *Clever Hans Phenomenon*, 1981.

372Terrace, *Nim*, p. 15, "Can an Ape Create a Sentence?" p. 892.

373Gardner and Gardner, "Comparative Psychology," pp. 321-25.

374Terrace, "Can an Ape Create a Sentence?" p. 900.

375Seidenberg and Petitto, "Ape Signing," pp. 118-19.

376Terrace, "Can an Ape Create a Sentence?" "Grammatical Capacity," "Reply," "Report."

377Seidenberg and Petitto, "Ape Signing"; Petitto and Seidenberg, "On the Evidence."

378Ibid.

379Petitto and Seidenberg, "On the Evidence," pp. 165-66.

[380] Ibid., p. 166.
[381] Patterson and Linden, *Education of Koko*, p. 141.
[382] Ibid.
[383] See discussion on the differences between gestures, iconic signs, and arbitrary signs in ASL, Petitto and Seidenberg, "On the Evidence," pp. 172-75—distinctions that Patterson and the Gardners appear to blur in their analyses.
[384] I.e., Patterson, "The Gestures of a Gorilla," pp. 72-97.
[385] Petitto and Seidenberg "On the Evidence," p. 168; see also the previous section.
[386] Patterson, "Linguistic Capabilities of a Lowland Gorilla," in F. Peng, ed., *Sign Language and Language Acquisition in Man and Ape*, A.A.A.S. Selected Symposium no. 16 (Boulder, Colo.: Westview, 1978), p. 178.
[387] Terrace, "Reply," p. 87.
[388] Patterson's doctoral dissertation, "Linguistic Capabilities of a Lowland Gorilla," Stanford University, 1979, p. 75.
[389] Terrace, *Nim*, chap. 13; idem, "Reply."
[390] Heini Heidiger, "The Clever Hans Phenomenon from an Animal Psychologist's Point of View," in Sebeok and Rosenthal, *The Clever Hans Phenomenon*, pp. 1-4.
[391] Ibid., p. 6.
[392] Ibid., p. 9.
[393] Ibid.
[394] Savage-Rumbaugh, Rumbaugh, and Boysen, "Do Apes Use Language?" p. 54.
[395] Bellugi and Klima, "Two Faces of Sign: Iconic and Abstract," in S. R. Harnad et al., *Origins and Evolution of Language and Speech* (New York: New York Academy of Sciences 280 (1976): 514-38, reported in Savage-Rumbaugh, Rumbaugh, and Boysen, "Linguistically Mediated Tool Use and Exchange by Chimpanzees (Pan Troglodytes)," in Sebeok and Umiker-Sebeok, eds., *Speaking of Apes*, p. 375.
[396] Heidiger, "The Clever Hans Phenomenon," p. 12.
[397] Sebeok and Umiker-Sebeok, "Questioning Apes," p. 22.
[398] Patterson and Linden, *Education of Koko*, p. 210.
[399] Chomsky, *Rules and Representations*, p. 239.

400. Donald Griffin, "Prospects for a Cognitive Ethology," *The Behavioral and Brain Sciences* 1 (1978): 527–38, quoted in Umiker-Sebeok and Sebeok, 1980, p. 14, n. 9.
401. Gardner and Gardner, "Comparative Psychology," pp. 317–20.
402. Patterson, "Linguistic Capabilities," p. 179.
403. Seidenberg and Petitto, "Ape Signing," p. 117; Petitto and Seidenberg, "On the Evidence," p. 169.
404. Gardner and Gardner, "Comparative Psychology," p. 319.
405. Patterson, "Gestures of a Gorilla," p. 179.
406. Seidenberg and Petitto, "Ape Signing," p. 117; Petitto and Seidenberg, "On the Evidence," p. 169.
407. Sebeok and Umiker-Sebeok, "Questioning Apes," pp. 36–41.
408. Ibid., p. 36.
409. Patterson, "Gestures of a Gorilla," p. 179.
410. Sebeok and Umiker-Sebeok, "Questioning Apes," p. 37.
411. Savage-Rumbaugh, Rumbaugh, and Boysen, "Do Apes Use Language?" p. 51.
412. Emil Menzel, "Natural Language of Young Chimpanzees," *New Scientist* 65 (1975): 130, as reported in Adrian Desmond's informative critique of the last decade of ape language research, *The Ape's Reflexion* (London: Blond and Briggs, 1979), pp. 231–32.
413. Menzel, "Natural Language," quoted in Desmond, *Ape's Reflexion*, p. 232.
414. Ibid., p. 231.
415. Terrace, "Is Problem-Solving Language?" (in Sebeok and Umiker-Sebeok, *Speaking of Apes*, pp. 385–405).
416. H. S. Terrace, *Intelligence in Apes and Man* (Hillsdale, N.J.: Erlbaum, 1976).
417. Ibid., p. 394 and passim.
418. See also Terrace, *Nim*, p. 27.
419. Terrace, "Problem-Solving," p. 394.
420. Ibid.
421. Ibid., p. 395; see Terrace, *Nim*, p. 28.
422. Terrace, "Problem-Solving," p. 392.
423. As reported in *Science News* (Dec. 5, 1981), p. 363.

Premack thought such training did make "some overall cognitive changes," however.

[424] Terrace, "Can an Ape Create a Sentence?" p. 899.

[425] Savage-Rumbaugh, Rumbaugh, and Boysen, "Do Apes Use Language?" p. 55.

[426] Claudia Thompson and Russell Church, "Explanation of the Language of a Chimpanzee," *Science* 208 (April 1980): 313–14.

[427] Ibid., p. 314.

[428] Lieberman, *On the Origins of Language*, chap. 9.

[429] Heidiger, "The Clever Hans Phenomenon," p. 9.

[430] Seidenberg and Petitto, "Ape Signing," p. 126.

[431] Savage-Rumbaugh, Rumbaugh, and Boysen, *Do Apes Use Language?*

[432] Ibid. They add a third presumption: "(3) The most important determinant of whether or not the chimpanzee is actually learning a language will be his ability to form novel combinations" (ibid.).

[433] Ibid., p. 52.

[434] See Chomsky, "Human Language," p. 434.

[435] Ibid., p. 437. Also see the numerous references on overattribution in The Interpretation Problem in this present work, pp. 127–34.

[436] Desmond, *The Ape's Reflexion*, p. 199.

[437] Terrace, "Can an Ape Create a Sentence?" p. 900.

[438] Chomsky, "Human Language," p. 437.

[439] Terrace et al., "Reply," p. 87; the authors take the data on Koko's M.L.U. from Patterson's dissertation, 1979.

[440] Terrace, *Nim*, p. 316.

[441] Terrace, "Can an Ape Create a Sentence?" p. 900.

[442] Seidenberg and Petitto, "Ape Signing," pp. 126–27.

[443] Sebeok and Umiker-Sebeok, "Questioning Apes," p. 21, n. 15.

[444] Eric Lenneberg, "A Neuropsychological Comparison between Man, Chimpanzee and Monkey," *Neuropsychologica* 13 (1975): 125.

[445] Akmajian, Demers, and Harnish, p. 344.

[446] Petitto and Seidenberg, "On the Evidence," p. 177, citing E. Klima and V. Bellugi, "Wit and Poetry in American Sign Language," *Sign Language Studies* 8 (1975): 203–24, citing E. Klima and V. Bellugi.

[447] For a detailed analysis, see Lynn A. Friedman, "Space, Time and Person Reference," in *Language,* 51 (December 1975): 940–61, especially pp. 951–52.

[448] Patterson and Linden, *Education of Koko,* chap. 21.

[449] Hediger, "The Clever Hans Phenomenon," p. 16.

[450] Terrace, "Can an Ape Create a Sentence?" p. 900.

[451] Seidenberg and Petitto, "Ape Signing," p. 122.

[452] Terrace, "Can an Ape Create a Sentence?" p. 900.

[453] Savage-Rumbaugh and Rumbaugh, "Symbolization," p. 278.

[454] Terrace, *Nim,* pp. 303–4.

[455] David Premack, personal communication to Adrian Desmond, November 15, 1978, recorded in *Ape's Reflexion,* p. 152.

[456] Chomsky, *Rules and Representations,* p. 57.

[457] Chomsky, *Aspects of the Theory of Syntax,* pp. 30–33.

[458] Chomsky, *Rules and Representations,* p. 239.

[459] James H. Stam, *Inquiries into the Origin of Language* (New York: Harper & Row, 1976), p. 245.

[460] Quoted in Stam, *Inquiries,* p. 245; from Charles Darwin's *Descent of Man,* p. 87 (New York, 1899).

[461] Lieberman, *On the Origins of Language,* pp. 22–24.

[462] Ibid., p. 21.

[463] Ibid.

[464] Ibid., p. 22.

[465] Ibid., p. 3.

[466] See Fromkin and Rodman, *An Introduction to Language,* chap. 2, for a simple overview of older naturalist views of human language development; Stam, *Inquiries,* chap. 4 for more detail; G. W. Hewes, "Language Origin Theories," in Duane Rumbaugh, ed. *Language Learning by a Chimpanzee: THE LANA PROJECT* (New York: Academic, 1977), pp. 3–53.

[467] Lieberman, *On the Origins of Language,* pp. 18–19.

[468] Patterson and Linden, *Education of Koko,* p. 204.

[469] Gardner and Gardner, "Comparative Psychology and Language Acquisition," p. 287.

[470] See chap. 8, p. 122.

[471] Chomsky, *Language and Mind,* pp. 66–71.

References

[472] Sir Karl Popper, Compton Lectures, n.d., reported in Chomsky, *Language and Mind*, pp. 67–68.

[473] Chomsky, *Language and Mind*, p. 68.

[474] Ibid.

[475] Ibid.

[476] W. H. Thorpe, "Animal Vocalization and Communication," ed. F. L. Darley, *Brain Mechanisms Underlying Speech and Language* (New York: Grune & Stratton, 1967) as cited in Chomsky, *Language and Mind*, p. 68.

[477] Ibid.

[478] Ibid., pp. 68–69.

[479] See chap. 3, pp. 41–42.

[480] Chomsky, *Language and Mind*, pp. 67, 70.

[481] Noam Chomsky, "Human Language and Other Semiotic Systems," in Sebeok and Umiker-Sebeok, *Speaking of Apes*, p. 438.

[482] Chomsky, *Language and Mind*, p. 70.

[483] Chomsky, "Human Language," p. 433.

[484] Noam Chomsky, discussion on "Properties of the Neuronal Network," in Piatelli-Palmarini, *Language and Learning*, p. 199.

[485] Noam Chomsky, "On Cognitive Structures and Their Development: A Reply to Piaget," in Piatelli-Palmarini, *Language and Learning*, pp. 35–36.

[486] Piaget, "The Psychogenesis of Knowledge," pp. 30–31.

[487] Chomsky, "On Cognitive Structures," p. 36.

[488] Ibid.

[489] Lieberman, *On the Origins of Language*, p. 19.

[490] Ibid., p. 3.

[491] Ibid., pp. 19–20.

[492] Ibid., pp. 5, 81.

[493] Ibid., p. 5.

[494] Ibid., p. 81.

[495] David Crystal, "Review of Philip Lieberman, *The Speech of Primates*" (The Hague: Mouton, 1972), in *Journal of Linguistics* 10 (September 1974): 330–33.

[496] Ibid., pp. 332–33.

[497] Noam Chomsky, "Language and the Mind," in Virginia P. Clark et al., eds., *Language: Introductory Readings*,

2nd ed. (New York: St. Martins, 1977), p. 339. Originally published in *Psychology Today* (Feb. 1968).

[498] Patterson and Linden, *Education of Koko*, p. 204.

[499] See Seidenberg and Petitto, "Ape Signing: Problems of Method and Interpretation," in Sebeok and Rosenthal, *Clever Hans Phenomenon*, footnote on p. 126.

[500] Daniel J. Bronstein, "Introduction," *Essential Works of Descartes*, pp. xi–xiii, xvii.

[501] Meditation III: "Concerning God: That He Exists," *Meditations on First Philosophy*, in *Essential Works of Descartes*, p. 80.

[502] Ibid., p. 81.

[503] Colin Brown, *Philosophy and the Christian Faith* (London: Inter-Varsity, 1969), p. 52. Brown remarks that Descartes was interested in God for the sake of the world, "to guarantee the validity of our thoughts about the world" (ibid.). Bronstein, "Introduction" to *Essential Works*, points out the historical context of Galileo's inquisition as having certain effect on Descartes's exposition of his ideas (pp. xii–xiv).

[504] *Discourse on Method*, Part V, in *Essential Works*, p. 34; also cited in Stam, *Inquiries into the Origin of Language*, p. 20.

[505] Falk, *Linguistics and Language*, pp. 8–9.

[506] Diane D. Bornstein, ed., *Readings in the Theory of Grammar* (Cambridge, Mass.: Winthrop, 1976), p. 5.

[507] Stam, *Inquiries*, p. 102.

[508] Ibid., pp. 118–27.

[509] Ibid., pp. 118–19.

[510] Ibid., p. 116.

[511] The general question of the origin of language has received considerable treatment from various sources down through the ages. See Stam, *Inquiries;* G. W. Hewes' survey "Language Origin Theories," in Rumbaugh, *Language Learning*.

[512] Mortimer J. Adler, *The Difference of Man and the Difference It Makes* (New York: Holt, Rinehart, and Winston, 1967), p. 17.

[513] From Stephen A. Grunlan and Marvin K. Mayers, *Cultural Anthropology: A Christian Perspective* (Grand Rapids: Zondervan, 1979), p. 267. Most of the comments offered in this response have been drawn from this book.

For Further Reading

GENERAL INTRODUCTION TO LINGUISTICS

Akmajian, Adrian; Demers, R. A.; and Harnish, R. M. *Linguistics: Introduction to Language and Communication.* Cambridge, Mass.: M.I.T. Press, 1979.

A good general introduction to language as a rule-governed system, concentrating on pragmatics (language usage in context) and semantics. It gives a good overview of animal systems of communication. It questions whether Sarah the chimp was able to form a sentence freely. A provocative text.

Lyons, John. *Language and Linguistics: An Introduction.* Cambridge, England: Cambridge University Press, 1981.

A good overview of the various views of grammatical theory. It has excellent sections on language and culture, sociolinguistics, and semantics.

TRANSFORMATIONAL GRAMMAR

Baker, C. S. *Introduction to Generative Transformational Syntax.* Englewood Cliffs, N.J.: Prentice-Hall, 1978.

A good book on syntax, covering material similar to the content of chapters 3 and 4.

Chomsky, Noam. *Language and Mind,* enlarged ed. New York: Harcourt Brace Jovanovich, 1972.

This book includes a review of the history of treatments on language and mind, from seventeenth-century rationalism to twentieth-century behaviorism. It expounds the new cognitive approach. It is especially valuable for Chomsky's argument, explaining the unfortunate historical change in emphasis from the Cartesian (rationalist) approach to the behaviorist view. It also expounds on certain problems for evolutionary theory brought to light by the newer cognitive approach. The enlarged edition adds three essays on the nature of language.

Chomsky, Noam. *Reflections on Language.* Buffalo, N.Y.: Pantheon, 1975.

This book is a semitechnical exposition on the general properties of language. Chomsky answers the criticism of the times in reviewing certain revisions of his "Standard" transformational theory.

LANGUAGE ACQUISITION

deVilliers, Jill, and deVilliers, Peter. *Language Acquisition.* Cambridge, Mass.: Harvard University Press, 1978.

A good survey of research on language acquisition through the late 70s. Fairly readable.

Dale, Phillip S. *Language Development, Structure and Function,* 2nd ed. New York: Holt, Rinehart & Winston, 1976.

This study is a bit older but is fairly readable. It contains less conceptual depth than deVilliers's; nevertheless, it is a good introduction to psycholinguistics.

Miller, George A. *Language and Speech.* San Francisco: W. H. Freeman, 1981.

Good on language and psycholinguistics, this text is a general survey of topics related to the origin, structure, and use of human language by a leading psycholinguist. The author holds to an evolutionary framework, but he has noted that at the present time we cannot account for certain things and has raised honest questions with respect to the language gap: "Human language requires so many special and anatomically unrelated adaptations that it seems unlikely they all appeared simultaneously in a single mutation. . . . If there were pre-linguistic forms of vocal communication, none of the hominids that used them have survived; there is today an enormous gap in communicative abilities between human beings and all other animals."

THE CHIMPANZEE STUDIES

Sebeok, T. A., and Umiker-Sebeok, J. *Speaking of Apes: A Critical Anthology of Two-Way Communication With Man.* New York: Plenum, 1980.

This is the best anthology, with the most important criticisms. It is not wholly critical.

Pike, Kenneth L. *With Heart and Mind.* Grand Rapids: Eerdmans.

This is a good attempt by the renowned linguist–Bible translator to integrate a scientific and a biblical view of language. (Out of print.)

HUMAN NATURE

Adler, Mortimer. *The Difference of Man and the Difference It Makes.* New York: Holt, Rinehart, and Winston, 1967.

Adler asserts that the difference between man and animals or computers must be qualitative rather than quantitative, though he does not state why. He suggests the necessity of a marriage of philosophy to the natural sciences. (Out of print.)

Cosgrove, Mark. *The Essence of Human Nature.* Grand Rapids: Zondervan/Probe, 1977.

Current evidence from psychological research is examined. It points to the inadequacy of those views of human nature that describe man as merely material, as totally determined, and as only a higher animal. A critical evaluation is made of recent brain-control experiments, the deterministic model of B. F. Skinner, and language studies in chimpanzees.

CULTURE AND LANGUAGE

Mayers, Marvin. *Christianity Confronts Culture.* Grand Rapids: Zondervan, 1974.

Mayers presents a case study of Christianity and its impact on world culture. He discusses and defines the impact of the Christian gospel, ethics, and life style—and of the one who introduces them—on one's own or foreign cultures. Numerous case studies. Bibliography and index.

Terrace, Herbert. *Nim: A Chimpanzee Who Learned Sign Language.* 1979. Reprint ed., New York: Washington Square, 1981.

This paperback is quite readable. Terrace reviews the four years of Project Nim. Terrace surveys the work of others and his own work (with numerous collaborators) at Columbia University and then records second thoughts about his work. His second thoughts make the book an important one.

Wilson, Clifford. *Monkeys Will Never Talk, or Will They?* San Diego: Master Books, 1978.

Dr. Wilson's earlier writing on a popular level. It includes some discussion of other animal communications such as bee dancing, birds' songs, dolphins' sounds, and pecking pigeons.

LANGUAGE ORIGINS

Stam, James H. *Inquiries Into the Origin of Language.* New York: Harper & Row, 1978.

Stam chronicles the progression of the discussion surrounding the origin of language, especially that of the Prussian academy in the late eighteenth and early nineteenth centuries. Since the whole issue has been reopened, this text provides some important historical background for the current controversy.

BRAIN RESEARCH

Custance, Arthur C. *The Mysterious Matter of Mind.* Grand Rapids: Zondervan/Probe, 1981.

Custance reviews various explanations that have been offered for the ascendancy of the mechanistic approach to the nature of the human brain and mind. He presents the experimental findings of recent research that have led some of the most renowned scientists in the field to conclude that mind is more than matter and more than a mere by-product of the brain.

Jones, Gareth. *Our Fragile Brains: A Christian Perspective on Brain Research.* Downer's Grove, Ill.: InterVarsity, 1980.

The author surveys the structure of the brain and research on its functioning, including split-brain studies and other phenomena such as the effect of psychedelic drugs.